TSUNAMI

THE WORLD'S MOST TERRIFYING NATURAL DISASTER

The proceeds from the sale of this book are going to CARE International UK. CARE International UK is an independent development organisation working to end world poverty. CARE International UK supports poor rural and urban communities to make positive and lasting changes to their lives. CARE International UK is part of the CARE International confederation, a worldwide humanitarian organisation which works in over 60 developing countries.

CARE International UK has not approved, endorsed or been in any way responsible for the contents of this book.

This is a Carlton Book

First published in Great Britain in 2005 by
Carlton Books Ltd
20 Mortimer Street
London W1T 3JW

ISBN 1 84442 411 1

10 9 8 7 6 5 4 3 2 1

A CIP catalogue record for this book is available from the British Library.

Printed and bound in Dubai

Managing Editor Lorna Russell
Edited by Justyn Barnes and Aubrey Ganguly at Aubergine
Production Lisa French
Art Director Clare Baggaley
Design Paul Sinclair at Parallax Studios
Maps Martin Brown
Diagrams Adam Wright
Picture Research Tom Wright

TSUNAMI

THE WORLD'S MOST TERRIFYING NATURAL DISASTER

Geoff Tibballs

CARLTON
BOOKS

CONTENTS

iNTRODUCTION

INTRODUCTION

Before 26 December 2004, many people had no idea what a tsunami was. And even the majority of those who were vaguely familiar with the term had little concept of the wholesale destruction that a tsunami could cause. To most people in the west, at least, tsunamis were the stuff of science fiction books, strange occurrences of centuries past, and, perhaps most importantly, things that happened on the other side of the world.

But all of that changed in a matter of a few minutes as a powerful earthquake deep in the Indian Ocean triggered the deadliest tsunami in history. Over a total period of seven hours and an area of more than 3,000 miles, towns and villages from Sumatra to Somalia were flattened as the giant waves destroyed everything in their path. At least 280,000 people were swept to their deaths, although the final death toll will never be known because of the countless bodies that were carried out to sea. But it will undoubtedly be considerably higher than the reported figure.

In the wake of the most devastating natural disaster of modern times, the world was united in grief. With so many tourists staying in the stricken regions over the holiday season, the tragedy was by no means confined to the countries lashed by the tsunami. Virtually every nation – from Argentina to Australia, Belgium to Brazil, Canada to China – has suffered some casualties, with the death tolls often running into the hundreds. The Indian Ocean tsunami was truly a global catastrophe.

The world has also been forced to re-evaluate its stance on the threat posed by tsunamis. Geologists have pinpointed a number of sites – in the Indian, Pacific and Atlantic oceans – where tsunamis may occur in the future. An Atlantic tsunami alone could endanger three continents – Europe, Africa and America – and would almost certainly cause widespread destruction, both to people and property. There may or may not be a major tsunami anywhere in the world for another hundred or more years, but any sense of complacency about this natural phenomenom has disappeared.

WHAT IS A
TSUNAMI? 1

A tsunami is a chain of fast moving waves (known as a 'train') that are generated when water in the ocean – or even a lake – is rapidly displaced by a sudden trauma in the form of an earthquake, a volcanic eruption, a landslide or the impact of a meteorite. Tsunamis are virtually undetectable out in the open sea – fishermen can be directly over a tsunami without experiencing more than a mild ripple in the water. However as tsunami waves race across the ocean at speeds of up to 600mph, they create havoc on reaching land, destroying everything in their path. As they approach land, the shallow water acts as a brake on the front of the wave, slowing it down to around 200mph, while the back of the wave continues at full speed. The back catches up with the front to form a terrifying wall of water. This marked contrast in behaviour at sea and near land gives the tsunami its name. For the term derives from the Japanese for 'harbour' ('tsu') and 'wave' ('nami'), and was coined by fishermen who returned to port to find the area surrounding the harbour devastated…

even though they had been unaware of anything untoward in the open sea.

Although often referred to as tidal waves, this is misleading as tsunamis are not caused by changes in the tides. They not only travel at great speeds across the ocean – faster than a jet airliner – but also they hardly lose any energy on the way. Tsunami wavelengths (the distance between wave crests) can be in excess of 60 miles (100 kilometres) and are sometimes separated by as much as an hour. Just when tsunami victims think the worst is over, even bigger second or third waves come along to wreak far greater destruction, the flooding sometimes extending over vast areas inland. Tsunamis do not necessarily make their final approach as a series of giant breaking waves. They may be more like a rapidly rising tide. This may be accompanied by considerable underwater turbulence, sucking people

under and tossing heavy objects around.

Professor Bill McGuire of the Benfield Hazard Research Centre at University College, London, explains the hidden menace of a tsunami: "With normal wind-driven waves that we see all the time, it's only the top of the sea that is moving but with a tsunami it's the entire ocean that is moving, from top to bottom. So when a tsunami reaches the coast, the level of water behind is just as high – it's like a wall of water that keeps coming."

The force of a tsunami can be such that rocks weighing 20 metric tonnes have been plucked from sea walls and carried 180 metres inland. Ships and boulders can be swept several miles inland before the tsunami subsides. In some instances, entire beaches have been swept away.

Another view of the destruction wreaked by the 1908 earthquake in Messina, Italy, which claimed the lives of 66,000 of the townspeople.

Parking meters in Hilo, Hawaii, uprooted by a tsunami that emanated from an earthquake off the coast of Chile.

The chaotic scene left behind in Anchorage, Alaska, after a huge tsunami struck in March 1964.

Geological features such as reefs, bays, river entrances and undersea formations may dissipate the energy of a tsunami. A beach guarded by a steep seabed may escape the ravages of a tsunami relatively unscathed whereas a neighbouring beach fronted by a shallow seabed will feel its full force and also encourage the water to spread inland.

Often the only warning of an impending tsunami is the sight of the tide suddenly going out as far as the horizon. This peculiar phenomenon has been compared to someone pulling the plug out of the ocean. Experts say that a receding ocean may give people as much as

five minutes' warning to escape to high ground. As the water is sucked out to sea, fish are exposed on the dry seabed but the eager fishermen or curious tourists who seek to investigate this freak of nature invariably pay with their lives when the water suddenly comes rushing back in as giant waves. In Thailand, in 2004, tourists were seen wandering around the exposed seafloor taking photographs, unaware that disaster was just moments away. Many of those who managed to survive the tsunami owed their safety to spotting the warning signs and reaching higher ground in time. As with the Lisbon earthquake and tsunami in the eighteenth century, animals seemed to have a sixth sense that danger was imminent. Many witnesses reported seeing animals fleeing for high ground minutes before the 2004 tsunami arrived. Very few animal bodies were found afterwards.

Japan has built a series of 4.5-metre-high (13.5 foot) tsunami walls to protect heavily populated coastal areas, but the tsunamis are often higher than the barriers. The tsunami that hit Hokkaido in 1993 created waves up to 30 metres (100 feet) tall – as high as a ten-storey building. Although the port town of Aonae was completely surrounded by a tsunami wall, the waves washed over the wall and destroyed all wooden-framed structures in the area. The wall may have slowed down the tsunami but it did not prevent major loss of life.

Sadly, even that paled in comparison to what took place in the Indian Ocean in December 2004.

The earth's crust is divided into giant rafts of rock called tectonic plates. These 12 individual plates are always moving, driven by the convection of heat from within the planet. At plate boundaries denser oceanic plates slip under continental plates in a process known as subduction. The India Plate is part of the huge Indo-Australian Plate, which underlies the Indian Ocean and the Bay of Bengal and is drifting north-east at an average of 6 centimetres (two inches) a year. The India Plate meets the Burma Plate (part of the Eurasian Plate) at the Sunda Trench off Sumatra. There, the India Plate has been slipping beneath the Burma Plate for millions of years, one plate pushing

A drawing of the island volcano of Krakatoa in Indonesia blowing itself out of the ocean in August 1883, generating a series of huge tsunami waves. The final one of four spectacular eruptions was reckoned to have made the loudest noise ever heard – louder even than an atomic bomb.

against the other until something has to give. Consequently the region is high in volcanic activity. Wherever plates meet, the risk of earthquakes is also great, and Sumatra lies at the extreme western edge of the Ring of Fire, a Pacific earthquake belt that accounts for over 80 per cent of the world's quakes.

At 00.58:53 GMT and 07.58:53 local time on 26 December 2004, the slow build-up of pressure caused by the constant grinding of the two plates resulted in a massive earthquake. Some 1200 kilometres (750 miles) of the faultline slipped about 15 metres (45 feet) along the subduction zone where the India Plate dives under the Burma Plate. This huge rapture is known as a

'megathrust' and produced an earthquake that lasted for more than four minutes and measured 9.0 on the Richter scale, making it one of the most violent on record. Seconds after the quake, the energy it released – the equivalent of 10,000 Hiroshima atomic bombs – was transferred to a water column above the ocean floor. In addition to the sideways movement between the plates, the seabed is estimated to have risen by several metres as a result of the quake, thereby displacing millions of tonnes of water above it. As the uplifted area collapsed, the water gushed away from it, creating devastating tsunami waves. Earthquakes can occur at depths of several hundred miles. When they do, much

An aerial view of the extensive structural damage caused to a section of coastal Chile by a tsunami in May 1960.

Houses in Castro, Chile, crumpled under 11-metre-high waves generated by the 1960 earthquake close to the country's coastline.

of their energy is absorbed by surrounding magma and rock. But this earthquake occurred at a depth of only 30 kilometres (18.6 miles), as a result of which there was no geological cushion. Had it happened much deeper in the ocean, it would not have generated a tsunami. As it was, the tsunamis radiated outwards along the entire length of the rupture, greatly increasing the area over which they impacted. Since the 1200 kilometres of faultline affected by the quake ran in a north-south direction, the tsunami waves were at their strongest from east to west. Thus Bangladesh, which lies at the northern end of the Bay of Bengal, sustained very few casualties despite being a low-lying country.

The water was soon powering towards coastlines all around the Indian Ocean. Within just 15 minutes of the earthquake, the tsunami had struck Aceh province in Sumatra, 155 miles north-east of the epicentre. 15 minutes later, it hit the Andaman and Nicobar islands and then Malaysia. An hour and a half after the quake, the tsunami hammered the beaches of Thailand, its arrival there delayed by the fact that the tsunami travelled more slowly in the shallow waters of the Andaman Sea off Thailand's west coast. Nearly two hours after the quake, the giant waves attacked Sri Lanka and the south-east coast of India. Next they lashed the Maldives, and, incredibly, seven hours after the earthquake, the tsunami savaged the coast of East Africa, 4,500 kilometres (2,800 miles) from the epicentre. Even at the end of such a long journey, it arrived with sufficient force to kill people and destroy property.

Past Tsunamis

Prior to December 2004 these were the major recorded incidents of tsunamis:

- **31 December 1703** The Genroku earthquake resulted in a tsunami that killed 100,000 people in Awa, Japan.

- **1 November 1755** When a colossal earthquake destroyed Lisbon, many Portuguese fled to the waterfront in the belief that the area would be safe from fires and falling masonry. Suddenly the waters retreated, revealing lost cargo and forgotten ship-wrecks, but it was merely the prelude to a deadly tsunami which hit the harbour just half an hour after the earthquake. Thousands who had survived the quake were killed by the six-metre-high (20 feet) waves, an estimated one-third of Lisbon's population of 275,000 perishing through the combined forces of the earthquake and the tsunami. Many animals are reported to have sensed danger and fled to higher ground shortly before the waters arrived.

- **1782** An estimated 40,000 people were killed by a tsunami that followed an earthquake in the South China Sea.

- **27 August 1883** In a series of four cataclysmic explosions that spewed grey ash around the world, the island volcano of Krakatoa in Indonesia blew itself out of the ocean, generating a series of huge tsunami waves over 40 metres (130 feet) high. As eleven cubic miles of rock was blasted into the atmosphere by the final, spectacular eruption, the sound – reckoned to be the loudest noise ever heard, louder even than an atomic bomb – resonated over one-twelfth of the earth's surface and was heard as far afield as Perth, Australia. On the facing coasts of Java and Sumatra, the flood from the tsunami extended many miles inland, drowning over 36,000 people. Waves from the eruption were recorded or observed throughout the Indian Ocean, the Pacific Ocean, the American West Coast, South America, and even in the English Channel. Coral blocks weighing 600 tonnes were hurled ashore, and a warship was carried three kilometres (two miles) inland by the waves and deposited on a low-lying hillside. It took just 12 hours for the tsunami to reach Aden on the southern tip of the Arabian Peninsula – a journey which, at that time, took 12 days by steamship.

- **15 June 1896** Spawned by an earthquake, the Sanriku tsunami unleashed waves as high as 30 metres (100 feet) on the east coast of Japan, killing around 27,000 people, including many who had gathered to celebrate a religious festival.

- **28 December 1908** At around 5.20 am, Europe's most powerful earthquake shook southern Italy. Centred in the Strait of Messina, which separates Sicily from the mainland, the quake triggered a tsunami, causing 13-metre-high (40 feet) waves to crash down on dozens of coastal cities. Many cities in southern Italy lost half of their residents that morning, the total death toll being estimated at 200,000.

- **1 April 1946** An earthquake on the Aleutian Islands off the coast of Alaska produced a tsunami that arrived at Hilo, Hawaii, some hours later, killing 159 people and causing millions of dollars in damage.

- **9 July 1958** Regarded as the largest of modern times, the tsunami in Lituya Bay, Alaska, was the result of an 8.3 magnitude earthquake followed by a landslide. Waves reached a height of 576 metres

The once thriving village of Sissano in Papua New Guinea was wiped out by a tsunami in July 1998.

(1,720 feet) in the bay, but because the area is relatively isolated and enclosed, damage was largely limited to the fjord itself. Indeed, this mighty tsunami sank a single boat, killing two fishermen.

■ **22 May 1960** A massive earthquake off the coast of Chile created a tsunami with 11-metre-high (36 feet) waves that killed an estimated 1,500 people in Chile, Hawaii, the Philippines and even Japan, 16,000 kilometres (10,000 miles) away, where it struck nearly 22 hours after the original quake.

■ **27 March 1964** The Good Friday earthquake in Alaska led to a 67-metre-high (210 feet) tsunami in the Valdez Inlet. Travelling at over 400mph, the waves killed 107 people in Alaska before rushing down the west coast of the United States and causing further deaths as far south as California.

■ **16 August 1976** A tsunami killed over 5,000 people in the Moro Gulf region of the Philippines.

■ **17 July 1998** An offshore quake trigged a 12-metre-high (40 feet) tsunami that struck the north coast of Papua New Guinea, killing some 2,200 people and leaving thousands more homeless.

MAIN
SHOCK
7.38 AM
MONDAY

← 12.09 PM

← 2.07 P.M

WATCHING
AND WAITING

As a result of the 1946 Aleutian Islands tsunami, which did so much damage hundreds of miles away in Hawaii, the countries of the Pacific Ocean established a tsunami warning system in 1949. At the Pacific Tsunami Warning Centre in Honolulu, geologists receive and analyse data provided by 120 seismometers located around the Pacific Basin. These seismometers measure ground motion and send signals as any movement occurs, thereby enabling any undersea earthquakes to be spotted instantly. Any ensuing tsunami is detected by a system of coastal tide gauges and deep ocean pressure sensors. However in December 2004 the Indian Ocean had no such warning facility, partly because the governments of the Indian Ocean nations could not afford it but also because it was not deemed necessary. Historically, the Pacific had always been at far greater risk from tsunamis. By contrast, the last major tsunami in the Indian Ocean had been caused by the eruption of Krakatoa back in 1883. Any threat from them appeared minimal…

A close-up of a seismograph chart recording the impulses of the Pacific Ocean's submarine earthquake which caused the 1946 Aleutian Islands tsunami and led to the establishment of the Pacific Tsunami Centre (pictured above, 28 December 2004) in Honolulu.

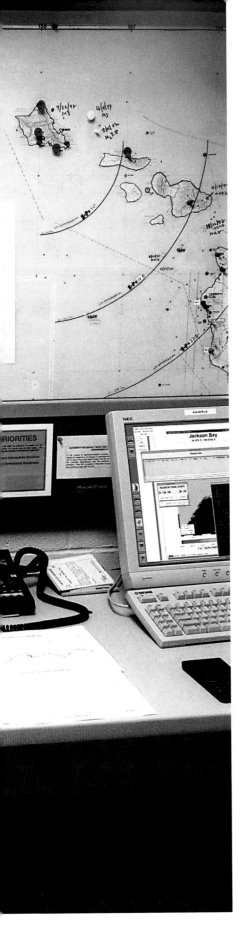

Indeed, although the governments in south-east Asia had actually been discussing the possibility of introducing an early warning system to deal with any threat from tsunamis just a year before the 2004 disaster, there was little support for the proposal because tsunamis were considered so rare. Whilst it may now appear to some to be a calamitous act of negligence, hindsight is a wonderful gift.

Barry Hirshorn, a geologist at the Pacific Tsunami Warning Centre, had been on duty for three days in the build-up to Christmas 2004 and was ready for a nap on the afternoon of Christmas Day (Honolulu Time). He was on the verge of falling asleep when his pager went off. The message told him of alerts from two far-apart seismic monitoring stations, meaning that whatever had occurred was big. He rushed to his office where his colleague Stuart Weinstein was on duty, already studying the thick blue seismic lines scrolling across the screen. For barely a minute after the earthquake had taken place on the other side of the world, the computers in Honolulu were picking up on the seismic signals.

The geologists' first task was to pinpoint the precise location of the earthquake and then to try and warn as many people as possible. Eighteen minutes after the quake they were able to issue an e-mail bulletin to 26 Pacific nations, announcing an 'event' off the northern coast of Sumatra with a magnitude of 8.0. However they realised that it was not the centre's Pacific nation clients that needed to be warned. The biggest hit would undoubtedly be in the Indian Ocean but because no warning system existed there, nobody in Honolulu had any indication of whether a tsunami had been generated. In fact, by the time that first bulletin was issued, the tsunami had already hit Banda Aceh on the northern coast of Sumatra.

Within the hour, although still unaware of the tsunami, the Hawaiian geologists had analysed sufficient quake data to issue a new bulletin upgrading it to a magnitude 8.5. Then via an Internet news report they learned for the first time about the tsunami, and revised the magnitude to a 9.0. Each upward revision indicated that the quake was several times more powerful than the geologists originally thought – a magnitude 9.0. quake being ten times stronger than one measuring 8.0.

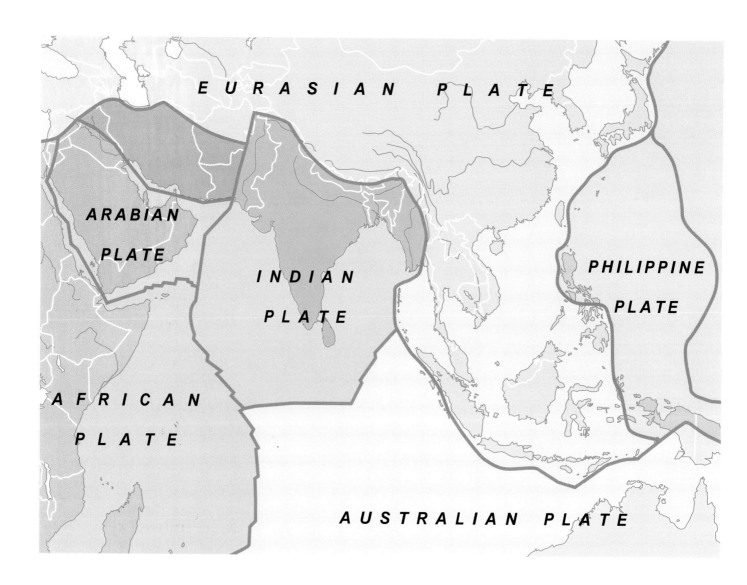

EURASIAN PLATE

ARABIAN PLATE

INDIAN PLATE

PHILIPPINE PLATE

AFRICAN PLATE

AUSTRALIAN PLATE

The earth's suface is made up of a series of plates. When they suddenly move together, an earthquake occurs.

It was now clear that a major disaster was on the way, but the frustrated scientists in Honolulu still had no way of alerting the potential victims – the tourists on remote island beaches, people living in coastal towns and fishing villages, anywhere from Indonesia to Africa. "We started thinking about who we could call," said Barry Hirshorn. "We talked to the State Department Operations Centre and to the military. We called embassies. We talked to the navy in Sri Lanka, any local government official we could get hold of. We were fairly careful about who we called because we wanted to call people who could actually help." The task was made more difficult by the fact that it was a holiday period in many countries and thus key officials were not readily available.

Obviously the priority was to contact people who were ahead of the wave. Stuart Weinstein created a tsunami travel time map for the Indian Ocean so that they could work out how much time they had to warn

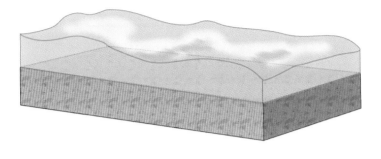

The tsunami in December 2004 was triggered by an earthquake deep in the Indian Ocean between the Australian and Eurasian plates.

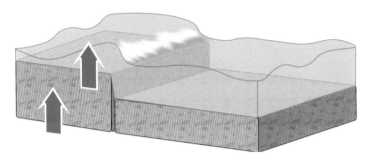

The energy from the earthquake vertically jolted the seabed by several metres, displacing hundreds of cubic kilometres of water.

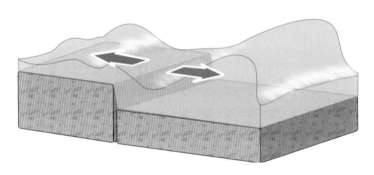

Giant waves began moving away from the earthquake's epicentre and the journey of the deadliest tsunami in history was underway.

people. Joined by their boss Dr. Charles McCreery, Hirshorn and Weinstein spent the next twelve hours on the phone, desperately trying to relay the urgency of the situation. Their sense of helplessness was compounded by the fact that even if government leaders could be reached, most of the smaller countries had no effective civil defence mechanism for passing on the information to the people at risk.

"We spoke to people in the foreign ministries, and everywhere we could think of," recalled Weinstein. "We were collecting phone numbers, e-mail addresses, whatever information we could. The message was simple: start walking away from the sea. You just have to be a 15-minute walk away from the sea to be safe."

Their valiant efforts may have been too late to save many countries, but

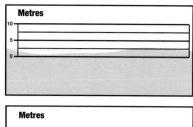

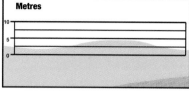

The tsunami moved at up to 800km/h (500mph) in deeper water. Shallow water acts as a brake on the front of the wave, but the back continues at full speed to form a wall of water.

the warnings did reach places like Kenya, Madagascar and Mauritius in time. Whereas the death toll in Somalia was high, only one person was reported to have been killed in Kenya because coastal areas were evacuated thanks to the advanced warning from Honolulu. Knowing that they had at least saved some lives helped ease a little of the frustration.

The 2004 Indian Ocean earthquake was more powerful than all of the world's quakes over the previous five years put together. It was the fifth largest quake ever recorded and the biggest for four decades. It was so big that it actually shifted the planet, knocking it off balance and shortening that day by three-millionths of a second. The total energy of the tsunami waves has been calculated at around five megatons of TNT, which is more than twice the total explosive energy used during the whole of World War Two. Some witnesses said the approaching tsunami sounded like three freight trains or the roar of a jet; others were taken completely by surprise.

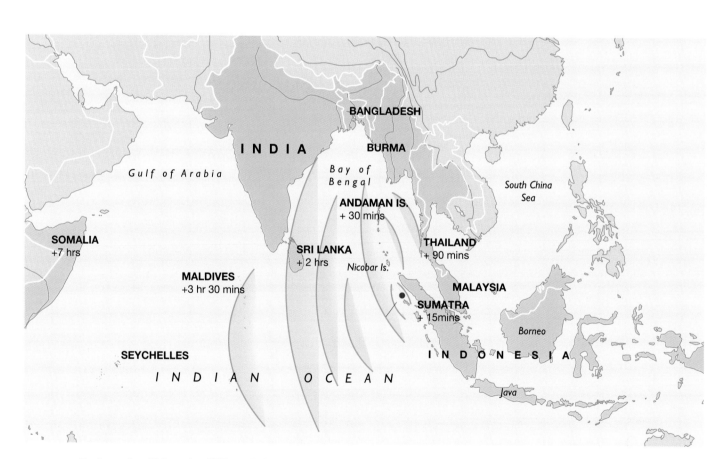

The tsunami on 26 December 2004 wrecked everything in it's path over a seven-hour period.

The incredible power of the tsunami was felt in countries across the 4,500 kilometre-wide Indian Ocean.

In some places the waves reached a height of 15 metres (50 feet), but elsewhere witnesses described a rapid surging of the ocean, more like a powerful river than the advance and retreat of giant waves. The force of the waves snatched people out to sea, drowned them in their homes or on beaches, or crushed them by hurling them against structures. Snorkellers were dragged across coral reefs, divers were trapped in caves, and fishermen were carried out to sea, never to return. Up to a third of the people who died in the Indian Ocean tsunami were children, many of whom would not have been strong enough to resist the power of the water.

Even adults were simply not able to run away fast enough. Survivors said that the sea surged out as fast and as powerfully as it came ashore with the result that many who had survived the incoming wall of water were seen being swept out to sea when the ocean retreated. Once caught in the torrents of raging waters, survival was down to pure chance. People who were together when the tsunami arrived were invariably separated. The fortunate ones managed to cling to buildings or trees; others were swept to their deaths. An eerie aspect of the disaster was that comparatively few people were injured. It was a case of death or survival.

Given the speed at which the tsunami spread, even an early warning system may not have saved the people of Sumatra. But it would surely have preserved the lives of thousands who perished in other countries further along the tsunami's path of terror.

INDONESIA 3

Pulau Breueh
Pulau We
Banda Aceh
KODYA
BANDA ACEH
Leupung
Seulimeum
Sigli
ACEH
BESAR
Meureudu
Gleebruk
Samalanga Bireun
PIDIE
Lhokseumawe Pantonlabu
ACEH UTARA Lhoksukon
Lhokkruet
Ibi
ACEH BARAT
Peureula
Calang
A C E H
Kualabee
Kualalangsa
Langsa Seruai
Kualasimpang
Meulaboh
ACEH SELATAN Blangkejeren
Lame
Tanjungpura
Belawan
Blangpidie
Labuhanhaji
Medan
Tapaktuan INDONESIA Tebingtinggi
Kandang
Sibigo
Trumon Sidikalang
Simeuluë
Sinabang
STRAIT
OF
MALACCA
INDIAN
OCEAN
S
U
M
A
T
R
A

INDONESIA

■ refugee camp
○ coastal settlements
● main towns
▢ worst affected areas

It was at eight o'clock on the morning of 26 December 2004 that the 225,000 inhabitants of Banda Aceh, capital of the Aceh province on the Indonesian island of Sumatra, had the first indication that something might be wrong. As the nearest big city to the epicentre of the massive undersea earthquake, Banda Aceh felt the tremors with greater force than any other built-up area…

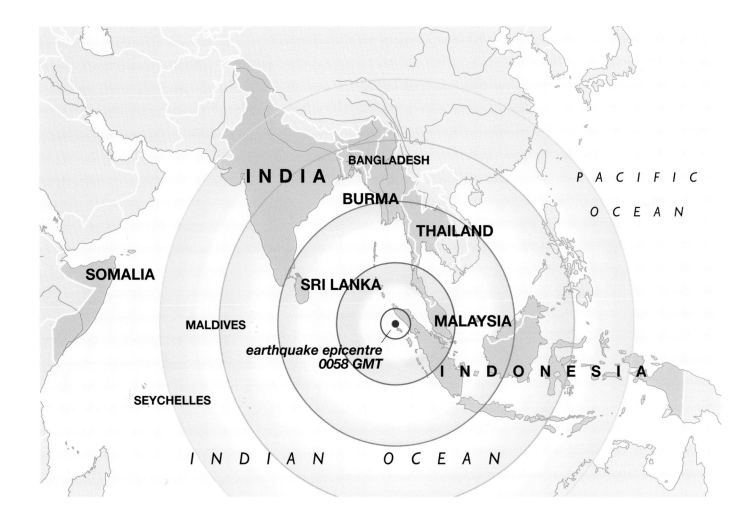

INDIA

BANGLADESH

BURMA

THAILAND

SOMALIA

SRI LANKA

MALAYSIA

MALDIVES

earthquake epicentre
0058 GMT

I N D O N E S I A

SEYCHELLES

P A C I F I C

O C E A N

I N D I A N   O C E A N

Some of the less substantial buildings wobbled precariously to the point of collapse, but out at sea all remained apparently calm. Fishermen quietly plied their trade, oblivious to the monster waves that were about to pound their homes.

The tsunami arrived some 15 minutes later. Witnesses described the wall of water as being twice the height of the coconut trees. At first as the waves encroached inland into the heart of the city, a cameraman, who was supposed to be taking wedding pictures that day, filmed what appeared to be a gentle river. People were still reasonably calm. Then the river suddenly turned into a raging torrent and widespread panic set in.

Mohammed Firdus had been sitting on the porch of his house, about 200 metres from the sea, when the earthquake struck. Then he heard a second rumbling, but this time the ground did not shake. "Someone came running fast from the beach, shouting, 'Huge wave, huge wave.' And then I saw the water. It was a wall at least a metre high coming down the track towards us. We all immediately turned and ran towards the main road."

CHAPTER OPENER:
Satellite images show the northern shore of Banda Aceh, capital of the Aceh province on the Indonesian island of Sumatra, before and after the devastation caused by the tsunami.

ABOVE:
Banda Aceh was the nearest large city to the epicentre of the massive undersea earthquake.

Holidaymaker Antonia Paradela was travelling on a ferry between the island of Pulau Weh and Banda Aceh when the tsunami hit. "I noticed that the ship bounced a little," she said, "but I didn't pay any attention to it. I was on the deck and an Indonesian guy pointed to the island, saying, 'Tsunami!' When I looked, I saw a wave that was going backwards; the spray was coming towards us instead of going towards the shore. I saw some people running on the shore but didn't realise the scale of it initially. In less than an hour we arrived in Banda Aceh and started to see that something out of the ordinary was happening. The jetties were submerged and some fishermen told us about a big wave."

Fearing that another tsunami was about to strike, the ferry captain headed straight back to the island but when Ms Paradela returned to Banda Aceh the following day, she could hardly believe the utter devastation caused by the tsunami. The city was a river of debris, with corpses floating down

Houses in a residential area of Banda Aceh were still submerged almost a month after the tsunami, 20 January 2005.

Men sit on a makeshift bench under the shade of a palm tree near the Tenom mosque in north-west Sumatra, 16 January 2005.

what were once bustling streets, people having drowned in their futile attempts to escape. With the city being built on low ground, the surging floodwaters were able to penetrate deep inland, washing away everything in their path. She said: "We saw big fishing ships that were thrown over buildings, and more and more bodies that lined the streets. The Chinese community area was in shambles. The cars appeared to be stuck to the walls, alongside the bodies of children and babies. The corpses were bloated and the majority of them had their arms upwards. The smell of the bodies was extremely pungent.'

Desperate relatives were invited to inspect piles of bodies lying in plastic tents. The stench of decomposing bodies hung in the air throughout Banda Aceh, even in the few parts of the city that escaped the worst of the tsunamis. Soldiers wearing rubber gloves and masks worked from dawn to dusk to try to clear the corpses in what was a truly thankless task. As the waters receded, bodies hung grotesquely from trees, and in every direction, human limbs protruded through the silt and mud. Given the number of casualties, it is hardly surprising that they soon ran out of body bags. With no proper equipment to lift the corpses, they were forced to use debris from the streets to lever the remains on to black plastic sheeting before trussing them up, as one correspondent put it, "like parcels ready for posting."

Despite the clear-up teams' finest efforts, five days after the disaster, countless corpses – many of them young children – remained strewn on the streets or floating in the rivers under a tropical sun. Two miles inland, a mound of around 200 corpses soon built up in the Kreung Aceh River under the Peunayong Bridge. To prevent the sanitation situation from worsening and fearing the spread of cholera and dysentery, government officials ordered that over one thousand bodies found on the streets of the provincial capital be placed in mass graves without waiting for identification. Many others were simply swept up into the mountains of debris that clogged the narrow thoroughfares, ready for future collection. And buried beneath the rubble and mud were thousands more bodies that are unlikely ever to be recovered.

Some 60 per cent of Banda Aceh was destroyed by the tsunami. The water left marks three metres up the adjacent hillside. A cement factory was wrecked, including a solid concrete wall that weighed hundreds of tonnes. Three miles inland, a 15-metre-long (50 feet) fishing boat was found washed up in the car park of a hotel.

Motorcyclists ride through the smoke and ruins of a street in Meulaboh, two weeks after the tsunami destroyed the city.

Banda Aceh was a city of 225,000 inhabitants with a highly-developed infrastructure...

Whereas many people – particularly children – did not understand the danger in the suddenly receding tide and made the fatal mistake of going down to the shore for a closer look, Hugh Crawford, an Englishman with a holiday home in a coastal village 11 miles outside Banda Aceh, recognised the signs all too well. Although he has only one leg, he knew it was imperative to get everyone up into the mountains as rapidly as possible. His fear was intensified by the size of the earthquake. "We heard the rumble and then felt the quake," he said. Dressing quickly, he led the exodus to the nearest high ground: "We watched the wave hit the harbour near the cement factory. But it didn't stop, it went up and on."

In the wake of the tsunami, Mr Crawford and other survivors decided it prudent not to return to the coast. They stayed on the mountain for three days before walking along the wasted shoreline because they had run out of coconuts and water, their sole source of food and drink. "It was really painful for me," he confessed,

... but even the sturdiest buildings were reduced to rubble as waves twice the height of coconut trees lashed the coast.

"because I'd run out of thick stump socks for my wooden leg and the skin was peeling. It was especially bad when we had to walk across the paddy fields because I'm not meant to get the leg wet."

The nearby town of Leupueng (population 10,000) was totally obliterated. Nothing vertical or square-edged was said to be standing, and the estimated number of survivors was put at no higher than a couple of hundred who managed to flee to the hills in the nick of time. It is thought that the coastal limestone cliffs directed the tsunami towards Leupueng. It was just one of a succession of small towns on the remnants of the coast road that ran south from Banda Aceh to Meulaboh. All were squeezed in between the sea and unforgiving rock faces, and therefore offered no protection whatsoever against the tsunami. Apart from the occasional palm tree, no landmarks remained to guide the few survivors in their search for relatives. One Leupueng survivor searching the rubble for his wife and two young children admitted: "I'm not sure what happened to my family.

It's mainly the men who survived because they can run faster." Another managed to locate the spot where his parents' house had once stood, only to find it completely destroyed. "I saw 11 bodies there," he said, 'but I have no idea if they were any of my relatives because their faces have disappeared."

Many who surveyed the terrible sight of the rubble that was Leupueng felt the same: how could a land that was once so beautiful have turned into this?

Meulaboh itself fared little better, being struck by a series of seven waves that killed at least half of the population of 120,000. At Calang, just north of Meulaboh, 90 per cent of the residents died – 6,550 out of a pre-tsunami population of 7,300. There, the trail of destruction left by the tsunami extended two kilometres inland. Entire hills were washed away. In West Aceh, the town of Teunom (population 18,000) was damaged so badly that, according to one report, 'it vanished completely leaving only scattered shards of concrete.'

At Kedah, a survivor described how the water came at his family from two directions, leaving them with only one escape route as the river blocked the other exit. "We all ran as fast as we could," he said, "but we couldn't get far. I was hanging on to my four-year-old daughter but the force of the water was so great she was ripped from my hands."

Prisoners in the town jail did not even have the chance to escape and perished where they stood. When the gruesome sight was discovered, many prisoners' hands were clenched into fists, as if they had been

People salvage what they can of their belongings in Tanah Pasir, Lhokweumawe, Indonesia, 27 December 2004.

The relief operation was hampered by damage to highways such as this in Leupung, 40 kilometres south-west of Banda Aceh.

banging on their cell walls when the tsunami struck. Even if anybody had heard their pleas, they would have been powerless to act.

Of 28 villages in the Lhoong sub-district, some 30 miles from Banda Aceh, only four remained. One survivor described how the earthquake lasted for around 15 minutes. "When that finished there was a noise like a bomb," he said. "Then a gap, then a noise like a ship's engine, and then people started screaming. Everyone started running with the water about 30 metres behind them. Many slipped and fell before they got to high ground. The water simply sucked them away and they disappeared. The water kept going for about five kilometres. Everything was destroyed. All the buildings were flattened."

As the water level dropped, water gathered in sinister-looking ponds. "The water was black in many areas," he added. "This probably means the ponds were full of bodies."

Fisherman Marwan Saad lived on a tiny island off the northern tip of Sumatra. "The earthquake didn't have that much effect," he said. "It just felt like a normal earthquake." However his calmness turned to fear less than 20 minutes later when villagers heard the sound of the approaching tsunami. "It was weird, like thunder," he said. "We were unsure about what it was.

INDONESIA

So we went outside and could see these huge waves coming from afar. Then we all started running, trying to find higher ground.' He carried his three-month-old daughter, while he and his wife dragged their two older children along. "The first wave destroyed my house, but the second wave was even bigger. I was submerged up to my neck and the children were completely underwater. I was forced to hold them as best I could." When the family returned to their house, they found nothing but the foundations.

Not only did the tsunami swamp the northern and western coastal areas of Sumatra and many smaller outlying islands, but also ten-metre waves passed the northern tip of the island and raced down the Straits of Malacca to strike along the north-east coast as far as the village of Lhokseumawe. In this fishing village, 150 kilometres from Banda Aceh, every house except for two was destroyed by the tsunami.

On the popular surfing island of Nias, off the west coast of Sumatra, an entire hotel, the Wismata Indah, was washed out to sea. The neighbouring island of Simeulue also lost 90 per cent of its coastal buildings but thankfully only five of the 70,000 villagers were reported to have been killed – and all of those perished as a result of the earthquake rather than the five-metre high wall of water that followed. The mayor explained their secret. "Thousands of our people were killed by a tsunami in 1907 and we have many earthquakes here. Our ancestors have a saying: if there is an earthquake, run for your life." In 2004, the villagers did exactly that and lived to tell the tale.

Owing to its geographical location so close to the earthquake, Indonesia suffered more than any other country at the hands of the tsunami. Over three-quarters of the total tsunami victims were from Indonesia, the vast majority from the devastated Aceh province. By the end of January the Indonesian death toll stood at well over 225,000 and was expected to rise significantly in the following months. In addition, the number of homeless people in Aceh and North Sumatra was estimated at a minimum of 800,000. In the immediate aftermath all infrastructure was wiped out in

A man sifts through the debris in Meulaboh, where a series of seven waves killed at least half the population of 120,000.

the worst-affected areas, leaving people without water, food or shelter. Fuel supplies were perilously low, forcing even ambulances to ration petrol and causing queues half a mile long outside petrol stations. Many local government officials, who might have been able to coordinate help, were either dead or missing. It was said to be impossible to calculate how many lives had been lost to the earthquake in Northern Sumatra because the tsunamis had followed on so swiftly to create far greater carnage.

Those who flew over the western Aceh coastline reported a scene of desolation. In many towns and villages concrete pads gave the only indication that substantial buildings had stood there just a few days previously, while scattered corrugated roofs, crumpled like paper, were all that remained of flimsier homes.

Occasionally a mosque stood intact among the brown wasteland, a defiant symbol at a time when prayer was the only salvation for so many. Roads and bridges had been swept away all along the western coast, increasing the difficulties faced by aid agencies in trying to get relief to the stricken settlements. Because so many areas were inaccessible, government officials were forced to make crude estimates of the death toll. Sometimes they would count the number of bodies in a mass grave and multiply it by the number of similar plots; in other instances they estimated the population of a village, counted the survivors and simply assumed that the remainder were dead.

A further complication in compiling an accurate picture of the devastation is the fact that due to the insurgency of the separatist Free Aceh Movement,

A large fishing boat dragged ashore and deposited in a residential area of Banda Aceh, 28 December 2004.

INDONESIA

In addition to the appalling death toll, by the end of January 2005, the number of homeless people in Aceh and North Sumatra was at least 800,000.

there were relatively few journalists or aid workers in Northern Sumatra prior to the tsunami. However on 27 December the government lifted the 18-month-old ban prohibiting foreign journalists and aid workers from travelling to Aceh, and the Free Aceh Movement declared a ceasefire so that humanitarian aid could reach survivors.

Yet from these appalling scenes came a few stories of miraculous survival. Two weeks after the tsunami had destroyed his home town of Calang, Ari Afrizal was rescued at sea, having lived on makeshift rafts and a leaky fishing boat. He and friends had been building a house in Calang when the wave struck. Initially he was pushed inland but like so many others he was then sucked out to sea when the tsunami waters receded. For 24 hours he clung to a log to survive. Then he clambered inside a damaged wooden boat before finally constructing a ramshackle raft from pieces of floating debris. "For three days I didn't get to eat anything," he said. "I gave up all hope of living. Eventually I managed to survive by eating the flesh of old coconuts."

He waved repeatedly at passing ships, but none stopped for him until an Arab container vessel appeared on the horizon. Even then, his attempts to alert the ship appeared to be in vain as it sped away. "I thought the ship had left the area," he said, "and I sat down and cried. But it returned for me."

Rizal Shahputra also survived at sea for eight days by clinging to a raft made of tree branches and debris. He had been cleaning a mosque in Banda Aceh when children ran in to warn him about the tsunami, but the surging waves swept them all out to sea before they could escape. He survived on rainwater and coconuts. "At first, there were some friends with me," he said,

"but after a few days they were gone. I saw bodies left and right. Everybody sank, my family members sank." He was eventually picked up after a ship spotted him floating 100 miles west of Aceh.

A young Indonesian woman, Malawati, suffered a similar experience. After being swept from her home, she nearly drowned twice because she could not swim, but while she was thrashing around in the water desperately trying to keep her head up, she chanced upon the trunk of a sago palm tree. She clung to it for the next five days, surviving by eating the fruit and bark of the tree. "I slipped twice but managed to hold on," she said. "I saw sharks around me but prayed they wouldn't hurt me." She was eventually picked up by a tuna ship and taken to hospital in Malaysia. Although badly bitten by fish and deeply traumatised, she had not lost the baby she was carrying. She did not tell her rescuers that she was pregnant because she

Destroyed fisherman's boats and houses in Meulaboh, 15 January 2005. The ravages of the tsunami extended two kilometres inland.

Rizal Shahputra miraculously survived for eight days by clinging to a makeshift raft before a container ship picked him up about 100 nautical miles from Aceh.

automatically assumed that she would have lost the baby in the ordeal but amazingly the hospital reported that the foetus was still alive and healthy.

Another heart-warming tale was that of nine-year-old Wira Dwi Lesmana. When the waves swamped Meulaboh, his family climbed a tree and scrambled on to the roof of their single-storey house. "But then Wira fell off and into the water," said his mother. "He climbed on to a cupboard but that soon began to sink too. He then clambered on to a mattress and I watched him drift out to sea. I was screaming and I heard him calling 'Mama, Mama.'" She feared the worst but two days later mother and son were reunited at the town's military base. He had somehow survived on the mattress without any food and water before climbing aboard a passing ferry.

Sadly these stories are very much the exceptions to the rule. Precious few in the coastal regions of Aceh province escaped unscathed. At best, they lost just their house, but most lost some members of their family as well as their home and livelihood.

SRI LANKA 4

INDIA

Palk Strait

Velanai I.
Jaffna
Chavakachcheri
Palk
Bay

Pamban Channel

Adam's Bridge

Mannar

Mullaittivu

Kokkilai

SRI LANKA

Gulf

of

Mannar

Vavuniya

Trincomalee

Kinniyai
Mutur

Anuradhapura

Kathiraveli

Puttalam
Lagoon
Puttalam

Polonnaruwa

Valachchenai

Mundel L.

Chilaw

SRI LANKA

Batticaloa

Kurunegala

INDIAN

Negombo

Kandy

Amparai

Kalmunai

Senanayake
Samudra

OCEAN

Colombo

Moratuwa

Ratnapura

Pottuvil

Kalutara

Tissamaharama

Hikkaduwa
Galle

Tangalla

Hambantota

Matara

- refugee camp
• main towns
☐ worst affected areas

The early morning train ferrying passengers from the capital of Sri Lanka, Colombo, 110 kilometres (75 miles) to the southern city of Galle was more crowded than usual because it was a Buddhist holiday when Buddhists offer special prayers and people visit relatives…

Some 1,600 passengers were crammed into the eight carriages or hanging from the sides as the train – named with grim irony *Samudradevi* or *Queen of the Sea* – trundled down the island's west coast, the track rarely veering more than a few yards from the sea shore. For much of the journey the sea is visible from the train, but at Telwatta the track cuts through thick palm groves, as a result of which the sea, although only 200 metres away, can barely be seen.

As the train stopped at signals, without warning a giant wave at least six metres (20 feet) high roared through the trees and swept the carriages off the rails. The tsunami had struck.

Shenth Ravindra from Crawley, Sussex, was a passenger on the *Queen Of*

CHAPTER OPENER:
The wheels ripped from the carriages of the *Queen Of The Sea* train which was carrying some 1,600 people from Colombo to Galle. It is estimated that less than 100 people survived.

ABOVE:
The twisted remains of the track.

The Sea. He remembered how shortly after the train had stopped, he heard screaming: "I looked out of the window and saw women running from the sea. A wave of water then flooded our carriage with a force strong enough to sever us from the rest of the train, carrying the carriage away from the tracks and tilting it at an angle."

His immediate thought was that a bomb had gone off; the idea of a tsunami never crossed his mind at that stage.

"I was one of the first to climb up on to the roof of the carriage. I still didn't know what was going on, I just wanted to get out of there."

Other passengers from the carriage followed and a sense of relief emerged that the worst was over. As people chatted about the derailment, the general consensus of opinion was that they had experienced nothing

Shocked villagers from Telwatta wander through the wreckage of houses and train carriages.

more than a freak wave. "The rest of the train was still on the track," said Mr Ravindra, "and the people inside seemed fine. They were seated and calmly waiting for help to arrive. I thought to myself, most of them probably haven't even got wet."

Just as he was cursing his misfortune at being in the only carriage to be thrown off the tracks, the second wave arrived – half an hour after the first. "I had been using a house in the distance to assess the water levels. Now I saw the water receding fast. When the second wave came, the horizon changed. All I could see was one enormous cliff face of water charging towards me. I thought that this second wave was going to toss me off the train and into the water but instead, miraculously, it neatly pushed the carriage towards a house behind us. A child was hanging on to me, so we jumped and climbed on to the roof of the house.

"I looked back and saw the carriage that had been next to ours floating in the water. It had been shunted off the tracks this time and had swivelled around 90 degrees. The water was choppy and the carriage was being

Rescue workers continue to remove bodies from the scene of the train crash caused by the tsunami.

SRI LANKA

Sri Lankan villagers wear scarves over their noses to cover the smell of dead bodies, 29 December 2004.

tossed around. I saw a large woman in a pink sari inside the carriage. She was moving back and forth with the momentum of the train. Then I realised that it was a dead body."

He then had to wade through a sea of dead bodies, gently pushing them to one side, before finally making his escape to higher ground.

Colombo restaurateur Daya Wijaya Gunawardana was another of the survivors. He recalled: "Suddenly the sea flooded through the train very high, very quick. The whole train was filled with water, and then it fell over." As the impact of the waves sent the carriages spinning, helpless passengers were tossed around inside. Mr Gunawardana was trapped for 45 minutes before he was able to escape. "I thought that we were killed," he said, "but we prayed to our God and because of that, I got up to a window and escaped. The whole thing was flooded, so everybody was trying to get out."

Israeli tourist Danny Shahaf was travelling on the train with a friend: "When my carriage was turned on its side by the waves, I started to panic. It was so quick and it washed us so far away. The carriage kept filling up

A satellite photograph taken on 12 April 2004 of the lush green landscape on the coastline of Gleebruk village in Sri Lanka...

with water. I was telling my friend to run to the front of the carriage because the windows there were still above water. I pushed my friend through the window to get her up and out of the carriage. There was a woman next to me holding her baby and trying to hold the window open with her other hand but as I tried to help her, the carriage filled completely and the water pushed the window shut. Only my friend managed to get out.

"Back at the other end of the carriage it was dark. I held my breath and thought, 'This is how you die.' As I thought that, the train flipped again, the water slid away and I waded towards the light."

Mr Shahaf managed to make his escape, but the

The same place, stripped of all vegetation and man-made structures by the tsunami, as photographed on 5 January 2005.

woman with the baby was not so lucky.

In total, over 1,500 passengers failed to emerge from the train, many drowning in the flooded carriages in what was the world's worst-ever railway disaster, the death toll being double that of the 1981 tragedy when a cyclone blew a train off a bridge and into a river in Bihar, India. The dead included local villagers who had scrambled aboard the *Queen of the Sea* during the interval between the giant waves. Others had sought refuge from the waters by huddling together beneath the carriages, only to be swept back out to sea. Three days after the disaster, rescuers called off the search for survivors and scores of bodies were buried in a mass grave.

'Even survivors paid a heavy price in coastal areas of Sri Lanka. Homes, crops and fishing boats were destroyed and at least 400,000 people lost their livelihood'

Buses at the central terminal in Galle, on the southern tip of Sri Lanka, were tossed around like toys by the power of the waves.

"It was the only thing we could do," explained the monk who performed a service to commemorate the dead of all faiths. "The bodies were rotting."

With so many people from the village of Telwatta dying when their homes were hit by the tsunami, it was difficult for officials to ascertain precisely which of the hundreds of corpses were those of passengers. But it is thought that probably no more than 50 people got out of the train alive.

Although the train disaster took place on the southwest coast of Sri Lanka, the east coast was actually the worst-hit area. With over 30,000 confirmed deaths, only Indonesia suffered a greater loss of life than Sri Lanka. About 1,200 were killed at Batticaloa in the east. At Trincomalee in the north-east, where the tsunami

reached more than 2 kilometres (1.25 miles) inland, 800 people were reported dead. In the neighbouring Amparai district, over 5,000 lives were lost. With the Bay of Bengal rarely hit by tsunamis, local people had flocked to the beach to watch the spectacular waves, only to be swept into the water. Other victims drowned after running to retrieve fish flung on to the beaches by the first giant waves. Many of the dead were children and the elderly, the least equipped to combat the awesome power of the tsunami. In addition, more than a million and a half people on the island – nearly ten per cent of the population – were rendered homeless by the disaster.

A group of unlikely beneficiaries from the natural disaster were 200 prisoners being held in a jail in Matara.

They escaped when waves swept away the walls. But mortuaries throughout the country were soon overflowing. In the town of Panadura, bodies spilled out into the sun from the hospital's eight refrigerated chambers.

Two hotels in low-lying Yala National Park that were popular retreats for the upper echelons of Colombo society were submerged by the floodwaters, as was a beach hotel near Trincomalee. The frail wooden shacks that hugged the west coast railway line along which the *Queen of the Sea* made her fateful journey were reduced to splinters. Standing barely 100 yards from the sea, they took the full force of the tsunami. Piles of discarded wood and asbestos roofing lay beside the rail tracks – all that was left of home for hundreds of Sri Lankans. Only a few concrete shops and garages survived the onslaught. Even the survivors paid a heavy price in coastal areas of Sri Lanka. Homes, crops and fishing boats were destroyed and at least 400,000 people lost their livelihood. In the coastal belt vast swathes of paddy fields were destroyed and the extensive salinization of land, a result of the saltwater flooding, rendered it unsuitable for cultivation. Furthermore, rubbish and debris, also deposited by the onrushing waves, littered these once prized agricultural regions. To add to the sense of helplessness, agricultural vehicles and equipment were destroyed, and canals and drains blocked. Underground sources of water were also salinated.

People sort through the ruins of their former homes in the south-western coastal town of Pettigaliwatta, 30 December 2004.

Other Sri Lankans earn their living in more precarious ways than working on the land. At the village of Kirinda, on the southern tip of the country, 175 deep-sea divers search the seabed daily for beautiful shells of all sizes and colours which eventually find their way to foreign markets to make bangles and jewellery. The divers work for four months a year – between November and February – and earn anything between $300 and $500 a month, which is good money by Sri Lankan standards. They earn extra money by taking foreign tourists deep-sea diving, charging them $100 a time, and spend the rest of the year fishing in the ocean.

Whereas many parts of the island were bathed in sunshine, on the morning of 26 December the skies were darkening over Kirinda. The deteriorating weather undoubtedly saved the lives of the divers, most of whom decided not to go into the sea that morning because there was insufficient light to guide them down to the ocean floor. Those that were loading up their equipment in the hope that the weather might improve were forced to run for their lives when they saw the onrushing waves chasing swimmers and bathers inland. They could only look on helplessly as the sea rushed in and swallowed their homes. Although the divers' youth and fitness enabled them to fend off the floodwaters and escape to higher ground, more than 100 people died in Kirinda and the majority of the houses in the picturesque village were razed to the ground. The savage sea may not have taken the divers' lives but in many cases it did take their equipment, the fishing boats they used to hire, and their homes, which invariably housed stockpiles of shells put aside while they waited for the prices to rise. And with no insurance to cover these losses, the future of the Sri Lankan seashell industry looked bleak.

The country's fourth largest city, Galle, was also badly battered, the low land allowing three mighty waves to penetrate up to one kilometre inland. In an all too familiar story, buildings near the sea were simply washed away. The town's fishing industry was decimated. Not only were the boats

A view of the damage around the hospital in Galle gives an indication of the huge rebuilding programme that is required.

shattered, but it was decreed that fish from the surrounding waters were unsafe to eat for the six months following the disaster because hundreds of bodies had contaminated the shallow sea.

Sri Lanka's tourist industry suffered similar hardships as a result of that dreadful morning. Overseas tourists enjoying the Christmas break swelled the numbers in resort hotels. But when the tsunami struck, they were forced to flee in the clothes they were wearing. Some did not even have that luxury and had to fly home wearing borrowed clothes. Passports and belongings were all abandoned to the rising waters. Not surprisingly many European tourists were unfamiliar with tsunamis and were happily taking photographs of the gigantic waves as they approached the beach, convinced that the water would turn back before reaching them. Some realised their mistake just in time; others were too late.

Whilst those on the beaches were at greatest risk, their hotels offered little protection against such enormous waves. Holidaymakers gazing out at the spectacle were stunned to find the waters suddenly crashing into their rooms, and had to run for their lives. As one British tourist put it: "We all thought, 'Ooh, big wave,' and only started to panic when it was just ten metres away."

There was no warning, although Debbie Bateson, a Berkshire physiotherapist staying at a hotel in Ahungala on the south-west coast of Sri Lanka, said she had been advised of the tsunami by a palmist the

A view along the coastal railway line in the southern Sri Lankan town of Lunawa, taken on the day of the tsunami.

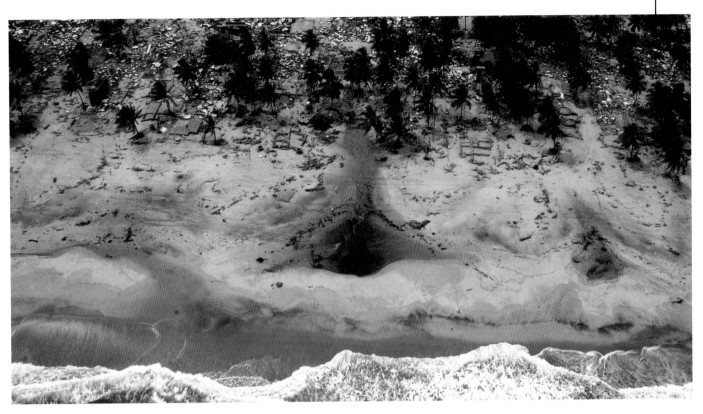

The denuded landscape along the southern coast of Sri Lanka as seen from a US Air Force helicopter.

previous day. "He said to me, 'Stay out of the sea – big wave.' I took no notice."

When the first wave engulfed the pool and entered the hotel foyer, she and her friend took the precaution of moving up to the third floor. Moments later a nine metre (30 foot) wave struck. "People were in total panic," she said. "It just came in within about a minute and a half. People were running and screaming."

In the wake of the disaster, tourists were evacuated from the eastern resorts to Colombo.

Michael Dobbs of the *Washington Post* was on holiday in Weligama. He was taking a morning swim in the sea when his brother shouted for him to come back because something strange was happening with the sea. "I couldn't understand what the fuss was about. There was barely a ripple in the sea. Then I noticed that the water around me was rising, climbing up the rock walls of the island with astonishing speed."

In less than a minute, the water level had risen at least 15 feet, but the sea remained calm with barely a wave in sight and the sky was still beautifully clear and blue. However within minutes the beach and the area behind it had become an inland sea that surged over the seafront road and into the flimsily-built houses on the other side. Instinctively Michael Dobbs and his brother jumped on a wooden fishing catamaran until the water stopped rising a few minutes later:

"I felt it was safe to swim to the shore, but what I did not realise was that the floodwaters would recede as quickly and dramatically as they had risen. All of a sudden, I found myself being swept out to sea with startling speed. Although I am a fairly strong swimmer, I was unable to withstand the current. The fishing boats around me had been torn from their moor-

Seawater flows through Galle
town centre, 26 December 2004.

ings, and were bobbing up and down furiously."

He swam towards a loose catamaran, grabbed the hull and managed
to pull himself to safety, eventually washing up on the sand. Across the
road, he heard screams coming from the houses, many of which were still
half-full of the water that had trapped the inhabitants inside.

One tourist who had an amazing escape was Welsh surfer Martin
Markwell. He was paddling on his surfboard off Hikkaduwa on Sri Lanka's
southern coast when he was swept up by a huge wave and sent crashing
over the beach into a hotel restaurant. Fortunately he managed to stay on
his board until he reached the hotel and, as the ocean rolled back to feed a
second much larger tsunami, he was able to jump off and wade to safety.
"I was surfing on a wave I wasn't supposed to be on," he said. "As an

experienced surfer, when I saw the wave come I realised something was wrong, but I couldn't escape because my surfboard was tied to my ankle."

Meanwhile an 11-year-old girl playing with coconut husks on the beach in the village of Pasikudha, near Batticaloa on the east coast, survived for more than 24 hours by grabbing a log when she was washed out to sea by the tsunami. During her ordeal, Sylvia Lucas had to fend off what she described as a "large fish" that was circling her. She saw people nearby being airlifted to safety by a rescue helicopter but she was too small to be seen. Finally the helicopter flew over again and she caught the rescuers' attention by waving frantically.

While Sylvia may appear to have been one of the lucky ones, it should be pointed out that she lost her brother to the waves. Alas, on 26 December 2004, that was the grim reality for many Sri Lankan survivors.

Welshman Martin Markwell, pictured with wife, Vicki, her son, Jai, and the surfboard that helped to save his life.

INDIA 5

Bombay
(Mumbai)

MAHARASHTRA

ORISSA

Western Ghats

Godavari

Vizianagaram
Vishakhapatnam

Hyderabad

ANDHRA PRADESH

Krishna

Vijayawada
Tenali
Machilipatnam

GOA

I N D I A

Ongole

KARNATAKA

Nellore

Eastern Ghats

INDIAN

Sira

OCEAN

Bangalore
Mangalore

Madras

Coromandel Coast

Mysore

Pondicherry

Malabar Coast

Salem

Cuddalore

Calicut

KERALA

Coimbatore

Kumbakonam

TAMIL
NADU

Thanjavur

Cochin

Madurai

Palk Strait

INDIA

Trivandrum

Gulf of
Mannar

Nagercoil

SRI
LANKA

- main towns
- worst affected areas

That Sunday morning, around 300 people had gathered on Manginapudi beach near the port of Machilipatnam on the southern coast of India to take a holy dip. The day was of particular significance as it was the eve of the full moon day in the Hindu holy month of Margasirsa...

Most had finished taking their dip but scores of worshippers were still innocently bathing in the sea when the tsunami swept in without warning. Several dozen people – among them elderly men and women and children – were snatched off their feet and carried out to sea. Their frail bodies stood little chance against the power of the waves.

Further up the coast, Christian and Hindu pilgrims flocked to the small seaside town of Vailankanni, known as the "Lourdes of the East", in search of cures at a popular shrine to the Virgin Mary. As many as 500 of these pilgrims died on the beach, entire families being wiped out.

As in Sri Lanka, the first tsunamis hit the southern shores of India around 9am, but it was the second – and sometimes third – waves nearly

CHAPTER OPENER:

Villagers move to higher ground in Port Blair, capital of India's Andaman and Nicobar Islands, on 30 December 2004 after the government issued what proved to be unfounded warnings of a second tsunami.

ABOVE:

The people of Chennai, the capital of Tamil Nadu, South India, salvage useful articles from the ruins of their homes.

an hour later that did the damage. Coastal communities in Kerala, Tamil Nadu, Pondicherry and Andhra Pradesh suffered a total of over 10,000 deaths with another 5,000 reported missing. Tamil Nadu was the worst-hit region, with some 8,000 lives lost. The tsunami struck Tamil Nadu with such ferocity that it blacked out a nuclear plant. Onrushing water flooded the cooling system of the Madras atomic power station, which was built to withstand cyclones and tidal waves, forcing an emergency shutdown of the reactor.

It was the first time that India had ever been hit by tsunamis, and people reported that they could still feel the aftershocks of the earthquake at 4.30 on that Sunday afternoon.

That the death toll was so high in India was attributable to the fact that the hamlets and villages perched on the edge of the sea – the buildings that took the full force of the waves – are the traditional preserves of impoverished fishermen. Their homes are not as sturdy as those of the rich

The destructive power of the Tsunami is graphically illustrated by the wrecked fishing boats along the Tamil Nadu coast. Approximately $125million will be needed to fix or replace the damaged vessels.

and middle classes who usually build their homes further inland, both out of choice and necessity since a law designed to protect coastal ecology prohibits any major construction within 500 metres of the shoreline.

Before the tsunami arrived, Nagappattinam was a bustling fishing town, its fleet of ten-metre trawlers and long, open boats the pride of the Tamil Nadu coast. Over 250,000 people in the region depended on fishing to earn a living. But the very sea that provided their income was about to take it all away. For within a matter of minutes the boats were smashed to pieces, sunk, or hurled hundreds of metres onto the shore. As the waters receded, vessels remained stuck in the gardens of houses or on railway tracks. They had been tossed around like toys in a bath. Officials estimated that of 15,000 fishing vessels along that stretch of coastline, only three escaped damage. A senior government official called the destruction of the fleet – and with it the men's livelihoods – "a catastrophe on a grand scale." He likened it to Delhi without its factories or the ruination of Bangalore's computer industry.

It was calculated that $125 million would be needed to fix or replace

A survivor in Nagappattinam sits quietly among the debris.

LEFT:
An elderly woman is helped to safety in Chennai, 26 December 2004.

RIGHT:
An aerial view of Nagappattinam post-tsunami. Twice as many women died as men here, because on Sunday most of the women were either at home or at the fish market by the sea.

damaged fishing vessels, buy new nets and lines and generally get the economic lifeblood of the coastal region flowing again. Remarkably, within two weeks of the disaster, a few boats had been repaired sufficiently to take to the waters once more. On their return to land, the fishermen reported that the tsunami – for all the misery it had caused – might actually have also brought new prosperity in terms of the vast numbers of fish now waiting offshore to be caught.

Although the economic hardship suffered by Nagappattinam was enormous, the human cost was even greater. Some two-thirds of the total number of dead in Tamil Nadu came from the area around Nagappattinam.

K. Panneerselvam was repairing his fishing net that morning, preparing to venture into the sea off Nagappattinam. He remembered the sea as being "calm and quiet." He was waiting for his colleagues to join him in their boats when he suddenly heard a roar. "There was no rain," he said, "and I immediately knew something was wrong."

When he looked out to sea, the vision before him was frightening. 'I have been fishing here 20 years, but I have never seen such huge waves.

Two days after the tsunami, stagnant seawater surrounds houses in Port Blair.

They were coming in like mountains. They hit me before I could turn around and scream out a warning to my wife and children." He managed to grip hold of the tree he was sitting under, but a few seconds later the tree came crashing down. "I thought I was dead," he said, "but I held on to the tree"' When he tried to get up, a second wave hit him. As he fell down, he saw his small, tiled house being washed away. Along with the house went his wife, a daughter and two sons.

In a classic tsunami scenario, fishermen told how they had headed 20 kilometres out to sea early in the morning, not noticed anything amiss, only to return in the afternoon to find their village had been devastated.

Close to the sea in Nagappattinam, humble homes and families were destroyed, twice as many women

dying as men because on a Sunday most of the women were either at home or at the fish market by the sea. Consequently a number of children were orphaned. Unable to find relatives or abandoned by a father who could no longer cope, they wandered aimlessly among the crushed houses or clung to familiar figures in their community. Refugee camps lined the roads and sprang up in the town's temples and colleges.

A local schoolteacher, Balakrishnan Ramadevi, told how a nine-year-old girl had lost her mother when the sea engulfed her house on the beach. Her grief-stricken father was so dazed and confused by the terrible events that he left the little girl to fend for herself. It was only when her teacher, Ms. Ramadevi, arrived at one of the refugee camps that the girl was identified. "As soon as she saw me, she came running towards me, crying, 'Miss, Miss,'" said the teacher. Of the 330 children at Ms. Ramadevi's school, only around 50 survived.

The refugee camps often provided contrasting emotions. On the one hand there would be parents grieving over their dead children, on the other there would be unexpected joy as people were reunited with family members whom they thought had been lost.

The business of burying the dead was clinical. One local man described how he and others pulled out 840 bodies over two days from flattened fishermen's houses in Nagappattinam: "If the body is in a condition to be moved, we put it into the mass burial pit and if it's too decomposed, we pour diesel over it and burn it with debris from thatched huts. Usually the pyres have 20 to 30 bodies at one go."

Clusters of fishing villages along the south-east coast of India were all but wiped out. At Kanaga Chettkuluan, Raja Shekar lost his livelihood, his family's possessions and ultimately his sister in just a few minutes. "I was working on my nets on the shore," he said. "We saw the water coming and I ran with my sister back to the house. The water came and it just dragged her back out. By then the water level was above the door." As others ran inland, he bravely climbed back into the house to search for his sister.

Nagappattinam locals gather around an upturned fishing trawler that was swept up onto a bridge by the force of the waves.

'Of the fleet of
15,000 fishing
vessels along the
Tamil Nadu
coastline, only three
escaped damage'

When the waters finally receded, he found her body, lying on the ground just 20 yards from their house.

Many fishermen stored their earnings in cash at home or spent their income on consumer goods such as televisions, mobile phones and compact disc players. The flood took it all. An added problem was the supply of drinking water. Before the tsunami, it was drawn from backwaters and local rivers, but contaminated seawater meant it was no longer potable.

Anton Raj, who lived in Kodimunai village, Tamil Nadu, was at work in the nearby town of Kanyakumari when the waves struck at around 9.45am. He drove home immediately to check that his parents were safe. On the way he passed three dead bodies on the road and came across hundreds of people wandering about, looking for transport: "In my village, 100 people were dead. Crying and inconsolable, people said only one thing: everything happened in five minutes. People said the waves did not make a noise until they hit the houses. First, water just came inside their homes but in just a few minutes, huge waves hit them and carried some away. Many old people died because they could not run. Couples who had more than two children also suffered because when you take two of your children, you can't leave a third or fourth one behind. The ones who could run fast managed to save their lives.

"My father told me that when the sea level was rising, he and many others actually went to see it! He said it was wonderful to gaze at the rising waters. People say that the sea was not arrogant, just silent. But within minutes, it rose to the height of a two-storey building and swept away their homes.'

The people of Cuddalore (just north of Nagappattinam) rush inland after another tsunami warning is issued via television, 30 December 2004.

A child stands amid the debris of ruined houses in Nagappattinam, 4 January 2005.

Navin Shetty and his friends arrived for a holiday in Pondicherry on the afternoon of 25 December. "On reaching the resort, which had eight cottages facing the sea, we found that it was not as good as the pictures we had seen on the Internet," he recalled. "Since the location was very good, we decided to spend the night in that resort and search for somewhere else the next morning." Consequently Mr Shetty was up early on the morning of 26 December. Had he still been in bed, he would surely have drowned.

He continued: "That morning, we were sitting in front of our cottage, planning the day. Around 9am, water started to surge inwards from the beach. Initially we were amused but then realised that the water had reached up to the compound wall. Before we knew it, a huge wave over six feet in height was upon us. We got worried and began to run outside the resort. When we reached the gate, we realised that one of our friends, who was not feeling too well, was still inside the room. I turned back, knocked on the door and called for her to come out. When she did not appear, I realised she must not have heard me and went inside the room. I saw her standing on the bed surrounded by water. She was afraid to come out, so I dragged her with me.

"By that time, the entire 200-foot-long compound wall had been razed, and water was all around the cottage. There was water up to my waist. I told my friend not to let go of my hand. We were moving slowly by holding on to the remains of the cottage but another wave raised the water level up to my neck, while my friend was completely submerged. We held on to each other's hand and headed towards a spot 15 feet away from our cottage, which was at a higher level. We had trouble moving because furniture was floating around

and banging into us. I could see our belongings floating in the water."

Eventually they reached the safety of the main gate and when the water started to recede, they returned to the cottage to retrieve what was left of their possessions. Mr Shetty added: "Although the water had begun receding, the waves continued, and each wave threw up dead bodies. I was there for just a few minutes but I must have seen between 20 and 25 bodies being thrown out of the sea.'

As is often the case with a major disaster, once the initial feeling of overwhelming despair has ebbed away, it is replaced by a sense of anger as people begin searching for answers. Inhabitants of the devastated regions of India demanded to know why the government had not warned them about the tsunami, given the length of time between the earthquake and the first giant waves washing ashore. Stung by the criticism and determined not to take any chances, the Indian government proceeded to cause utter chaos four days later on 30 December by issuing what proved to be

Following a gas cylinder exploding and catching fire, Indian firefighters go to work in the village of Malacca, close to Car Nicobar island, 2 January 2005.

unfounded warnings of new tsunamis. Thousands fled the south-east coast with what little remained of their possessions and fled inland. Meanwhile the army took to the streets to prevent anyone from getting within two kilometres (1.25 miles) of the coastline. The warning halted the recovery of bodies on mainland India and sparked panic in Port Blair, capital of the Andaman and Nicobar Islands, where residents fled the streets and many drove to the airport in an attempt to escape. In Sri Lanka, coastal villagers climbed on to rooftops or sought high ground and in Thailand sirens sent people rushing from the beaches.

Eventually Indian ministers took to the airwaves to retract previous warnings. There were no new killer waves. They had simply over-reacted to a string of underwater earthquakes near Sumatra that had been recorded by Hong Kong's observatory. But the quakes were 1,000 times less powerful than that of 26 December. However in the climate of fear that existed at the time, nobody was willing to take any chances.

A bustling city scene as the residents of Port Blair begin to get back into their normal daily routines, 3 January 2005.

THAILAND 6

Bada
Lord Loughborough I.
B U R M A
Chumphon
Kra Buri
Ban Sawi
Zadetkale Kyun
(St Lukes I.)
Laun
Isthmus
of Kra
Ranong
Zadetkyi Kyun
(St Matthew's I.)
RANONG
Kapoe
Ko Phangan
Khao Lang
Kha Toek
Chaiya
Ko Samui
Ko Phra
Thong
Surat Thani
Takua Pa
T H A I L A N D
G U L F O F
Khao Lak
Thap Put
PHANGNA
Phangnga
Nakhon
Si Thammarat
T H A I L A N D
Ban Khok Kloi
Ao Luk
KRABI
Thalang
Ko
Yao Yai
Krabi
Thung Song
Khao Chum Thong
Ko Phuket
Phuket
TRANG
Ranot
A N D A M A N
Ko Lanta
Lanta
Trang
Phatthalung
*Thale
Luang*
Phi Phi
Ban Kantang
S E A
Ko Libong
Suso
Songkhla
Hat Yai

• main towns

☐ worst affected areas

THAILAND

On the holiday paradise of southern Thailand tourists from across the globe were preparing for another day of minimal exertion. Some were already on the beaches or taking a morning dip in the sea, while the tardier were still labouring over breakfast in their hotels or apartments...

The resort of Phuket is at the heart of Thailand's tourist industry, supported by beautiful island beaches such as Phi Phi and Krabi.On the morning of 26 December, a few of the tourists were still nursing hangovers, having celebrated the night away in idyllic beachfront bars on a warm Christmas evening far removed from the wind-chill factor of their homes in Britain, Sweden or Germany.

Tilly Smith, a ten-year-old schoolgirl from Surrey, was too young to drink, which was probably just as well since it meant that she had her wits about on her that gloriously sunny morning. Like everyone else on the beaches, she knew nothing of the momentous earthquake off Sumatra or of the tsunami that was fanning out across the Indian Ocean and heading directly for Thailand. Fortunately she did remember a recent school geography lesson in which her teacher, Mr Kearney, had taught the class about earthquakes and how they can cause tsunamis. So when the sea off Maikhao Beach, Phuket, suddenly rushed out and started to bubble, and boats on the horizon began to bob up and down with greater force, Tilly

The resort of Phuket is a popular winter holiday destination for European tourists looking to escape the chill of their home countries...

... but on the 26 December 2004, they witnessed the extraordinary damage caused by a tsunami.

froze with fear. Her parents and other holidaymakers were simply curious but Tilly saw the danger signs. "Mummy, we must get off the beach now!" she said. "I think there's going to be a tsunami." The adults still did not grasp the severity of the situation until she spelled out the magic words: a tidal wave.

Her warning spread like wildfire and within seconds the beach was deserted. The Smiths retreated to the first floor of their hotel and minutes later the water surged right over the beach, demolishing everything in its path. Thanks to Tilly, Maikhao turned out to be one of the few places along the shores of Phuket where nobody was killed.

Over 5,000 people – half of them foreign holidaymakers – were less fortunate as the tsunami lashed the western coast of Southern Thailand. Not only was there a train of waves, but they came in rapid succession – just a couple of minutes apart – to sweep up young and old alike. Most witnesses spoke of three main waves, each deeper than the one before. The first wave knocked them off their feet; the second picked them up and carried them at speeds of 30mph; and the third, the most powerful,

This sequence shows Phuket's Chedi resort staff preparing for the day as the first swell of the tsunami edges the lawn...

... followed by the arrival of the second and third waves which engulf the hotel and the surrounding area.

either lifted them high into the air or sucked them under. Although the island of Phuket suffered extensive damage, its modern high-rise concrete hotels did offer some protection against the waves, enabling people to escape the raging floodwaters by climbing to the upper floors. However that option was denied to most other areas. Whereas only a few hundred died in Phuket, in the area around the smaller but increasingly popular resort of Khao Lak, some 80 kilometres (50 miles) north of Phuket, over 4,000 people were killed, principally because the accommodation in Khao Lak was largely composed of bungalows. Furthermore, most of the bungalows were built on an extensive area of flatland, just a few metres above sea level, leaving them at the mercy of the tsunami. At least 3,000

Beneath blue skies, the roof of a displaced gift shop is visible close to the beach at Phi Phi Island, 27 December 2004.

people died on a single short stretch of beach.

At one hotel in Khao Lak, around half of the 400 guests died in their ground-floor rooms, swamped by the force of the waves. Children playing in the lobby were swept away. When the waters finally receded, the blossoming resort looked like a bombsite. Luxury hotels lay in ruins, roofs collapsed on dining rooms, swimming pools filled with mud.

German tourist Jürgen Hegel owed his life to his refusal to wear sunblock during a Christmas Day game of football on Khao Lak Beach. He was so badly burnt as a result of playing in the game that the following morning he decided to keep out of the sun by going for a walk in the forest. He set off inland after breakfast and when the tsunami struck he was about a quarter of a mile from the beach. "I heard the roar of the sea, the

Clean-up crews work among what used to be shoreline holiday bungalows on Patong beach.

These images taken by Australian tourist Graham Francis with his cellphone shows the tsunami hitting Patong beach.

screams of the people on the beach and then the crash, crash, crash as the water smashed into the coast and everything in its path," he said. "Then I saw the water coming through the trees, so I turned and ran as fast as I could until I got to slightly higher land."

Mr Hegel and several others stayed on the small hill for several hours before daring to venture back down. "The return journey was an obstacle course of uprooted trees, boats that had been washed ashore, concrete from buildings, glass from smashed windows, and bodies. I must have passed well over 100 bodies. They were slammed against trees, wrapped around each other as if hugging, and buried under blocks of concrete."

Holidaymakers on Koh Ngai described how people who were snorkelling were dragged along the coral and washed up on the beach. Meanwhile sunbathers were swept out to sea.

British holidaymaker David Quinn was walking with his mother along the promenade from their Phuket hotel to Patong Beach. He said: "We had got as far as the Kodak shop to put our films in to be developed when we saw the crowds looking out to sea. In a moment everyone started screaming. There was a huge wall of water coming towards the beach. We turned and ran towards our hotel and kept running. Everything was getting washed away or smashed up. I saw people sucked under cars – it was horrendous." They made their way to higher ground and stayed in the mountains for the rest of the day as they had been told that another giant wave was coming. When they eventually returned to the beach area, everything was devastated. "We saw what was left of the Kodak shop: there was a bus on top of it. We would have been dead if we had gone into that shop – we had been seconds away from death."

On the two Phi Phi Islands, where the movie *The Beach* was filmed, hundreds of bungalows were washed out to sea. Charlie Anderson, a British tourist, was snorkelling from a boat off Phi Phi when the first wave came at around 9.30am. "It was like being at the foot of a mountain of water," he recalled. "The boat was simultaneously being lifted by the wave and I remember being pulled backwards and the propeller passing very close to my face. Then before I knew it, the wave must have passed over me and I was in the undercurrent, being sucked down at such incredible speed right to the bottom where I was then spun around. The third wave took me all the way to the bottom and I ran out of breath, so I basically died under the water. I remember physically struggling, then I breathed in and swallowed water before passing out. Luckily the fourth wave, which wasn't as big as the others, washed me on to the beach."

When the waters subsided from Phi Phi, they left behind a macabre scene. A dead shopkeeper was found wedged behind her counter while bodies stood pinned against walls, arms outstretched. In some cases their eyes were still open and they were standing up as if frozen to the spot, their hands raised as they cowered from the waves.

Those who escaped the tsunami on Phi Phi headed for a nearby hill where they spent several hours waiting for the situation to ease. Elsewhere drivers of tuk-tuks (Thailand's motorised rickshaws) were quick to offer assistance, driving victims to hospital or to high ground away from the surging waters. On one beach in Thailand, a man was leading an elephant to entertain tourists, when the tsunami came. By putting several children on the elephant's back, he saved them from the flood. Elephants also played a part in the subsequent rescue operation, being employed to lift heavy wreckage in order to facilitate the search for victims. With no hope of finding survivors in the rubble of crushed

A tourist hotspot turned temporary garbage dump close to Patong beach, Phuket, 30 December 2004.

The luxurious Sofitel hotel resort in Khao Lak, Phuket, in all its glory before December's tsunami…

hotels, bulldozers, tractors and sniffer dogs were brought in to help locate bodies covered by debris. While corpses were still being washed up on Thai beaches several days after the waves struck, many of the estimated 8,000 injured were evacuated to cities on the untouched eastern side of the peninsula.

Of all the dreadful images that flashed around the world, one of the most poignant was of a desperate woman running into the wall of waves on Krabi Beach as other tourists fled for their lives in the opposite direction. She was clearly searching for her family but it was naturally assumed that she must have died in the quest. Happily, a few days later it was revealed that the woman – 37-year-old Swedish policewoman Karin Svaerd – had indeed survived. "I can remember the white foam, how the surf took up my husband and three children, and they disappeared," she explained later. "I could hear people shouting at me to get off the beach as I ran past them, but I ignored them. I had to try and save my children. Nothing was going to stop me." She said she thought she would die when she was engulfed by the tide, but in fact it swept her onto higher ground and she managed to grab hold of a palm tree. Nevertheless she had to wait an agonising ten minutes before discovering that her family were safe too.

On every beach that morning people were faced with making instant decisions that could, without any exaggeration, prove the difference between life and death. Few were more agonising than that which confronted Australian holidaymaker Jillian Searle who was staying in Phuket with her husband Brad and sons, Lachie, five, and Blake, aged two. The tsunami struck as she and the children were strolling past their hotel pool. With her husband looking on helplessly from the family's first-floor room, she grabbed the children as the waters rose alarmingly but soon realised that she would not be able to hold them both. "I knew I had to let go of one of them," she said, "and I just thought I'd better let go of the one that's the oldest." So she let Lachie go and was relieved to see him caught by another woman, but unfortunately the woman was unable to hang on to him. For two hours the Searle family searched the resort, screaming for their lost son.

Mercifully, he was eventually found clinging to a door, deep in shock, but uninjured.

Australian photographer John Russell was standing on the beach at a diving shop in Koh Raya, just south of Phuket, when the tsunami struck. "There was no loud noise or wind," he said, "just all of a sudden the sea had risen a good ten metres. Then the wave suddenly sucked away from the shore a few hundred metres, exposing the coral reef that I had dived on a few days earlier. That's when people realised something was terribly wrong." He told how dive instructors and staff from a nearby restaurant ran down to the beach to see if anyone had been carried out to sea, only to find

... the scene of destruction at the same resort, 6 January 2005.

"There was a huge wall of water coming towards the beach. We turned and ran towards our hotel and kept running"

themselves faced with a surging ocean at least 15 metres high. When the second wave came in, it simply tore apart the wooden buildings that sat at the top of the beach. Then as that wave sucked out, everyone headed for the slopes on the sides of the bay. He added: "The third and probably largest wave then came surging forward and ripped apart the cement buildings like they were made of balsa wood." Mr Russell said the waves continued for about an hour, gradually decreasing in size before revealing the full extent of the damage.

Australian honeymooner Dave Ali was in his hotel at Phuton Beach, Phuket, when waves suddenly swept into the building: "There was a couple standing in the foyer just checking in when this thing hit – and that was it; that was their holiday – just wrecked straight away. There was one poor mother's baby got pulled out of her arms. They never found the baby. There was a father who couldn't find his wife or son – he ended up with broken ribs. We finally found his wife but his son was never retrieved. And a lady in a wheelchair just couldn't get away. I think she got swept away."

Local people also suffered terrible losses, notably in the fishing villages around Khao Lak. The fishing port of Tablamu was reduced to debris, boats being hurled ashore where they jostled for space with upturned police cars. Further north at Ban Namkhem, almost every house was flattened.

In the wake of the disaster, anxious families flew in to Thailand from all over the world for news of loved ones. For many it was a long, painful trawl of embassies, hospitals and, all too frequently, morgues.

Two rescuers carry away another body recovered from the wrecked Khao Lak Laguna Resort Hotel, 29 December 2004.

THAILAND

A man steps over an uprooted tree on his way through the remains of a hotel resort in Khao Lak in the Phang Nga province, 9 January 2005.

On arrival at morgues, relatives were warned by officials that the bodies were virtually unrecognisable due to the bloating of the water and the decomposition in the heat. Forensic experts were sent out by some countries to assist with the process of identification. Photographs of missing persons were plastered over Thai resorts and in hospitals in the hope of jogging memories, and pictures of unidentified survivors were posted on the Internet. As a result a two-year-old-boy found sitting alone on a road in the devastated town of Khao Lak was reunited with his family after an aunt in Sweden recognised his picture. Viola Hellstroem was scouring the web for news of her holidaying family when she stumbled across a photograph of her little nephew, Hannes Bergstroem. The picture of the boy, unharmed and smiling in the arms of a nurse, had been posted on a special web page set up by Phuket International Hospital, after staff had been unable to identify him.

The bodies from Phi Phi were taken to a makeshift morgue in a Krabi temple with up to 100 a day arriving in the early stages. At first the authorities allowed relatives to view the corpses in the quest for identification but, with no refrigeration and the bodies becoming increasingly bloated, this practice was stopped. Instead relatives had the gruesome task of looking through hundreds of photographs of the corpses that had been washed ashore, with notes about identifying marks, such as tattoos or jewellery, being kept at the hospital for families to study.

Even though many more bodies undoubtedly lay buried under the rubble, the Thai authorities were eager to begin the clean-up operation and bulldozers were brought in to shovel the debris into massive pyres

to be burned. Extra police were also drafted in to deal with an outbreak of thefts from corpses as well as with looting from wrecked shops.

Anger was not only directed at the thieves but also at the government's failure to warn of the tsunami. A former weather forecaster at the Thai Meteorological Department claimed: "The department had up to an hour to announce the emergency message and evacuate people, but they failed to do so. It is true that an earthquake is unpredictable but a tsunami, which occurs after an earthquake, is predictable." Instead warnings of a possible undertow on beaches issued by the department were apparently broadcast on television and radio after the first waves hit.

As the country tried to come to terms with its loss, it was the plight of the orphaned children that captured hearts and sparked renewed

Swedish policewoman Karin Svaerd runs towards her family as the first of the tsunami waves rumble towards the beach in Krabi, Southern Thailand.

determination. Bhumibol Adulyadej, the King of Thailand, whose 21-year-old grandson, Bhumi Jensen, was swept to his death while riding a jet-ski, reportedly promised to pay for the education of every child who lost a parent in the disaster. And the nation warmed to a 20-day-old Thai boy who somehow survived the tsunami that apparently killed both his parents, but was then abandoned three days later by the carer who was too poor to look after him. He was found in a bag in a national park on 29 December. The attached note read: 'Please, please look after this baby because I have no money and no means to keep him. His parents were swept into the sea at Patong. If you cannot keep him yourself, please take him somewhere where he will be looked after.'

The little boy, remarkably well considering his ordeal, was taken to hospital in Phuket. The nurses named him Wave.

FROM LEFT: Husband Lars, sons Viktor, Anton and Filip and their brave mother Karin Svaerd safely at home in Sweden on New Year's Day 2005 after their miraculous escape from the waves.

THE NET OF DESTRUCTION

Nearly 3,000 miles from the epicentre of the earthquake, the tsunami had one last chilling act to perform. Seven hours after the waves had started to spread out across the Indian Ocean at ferocious speed, they hit Somalia on the east coast of Africa, killing at least 150 people, leaving thousands homeless and wrecking the livelihoods of as many as 30,000.

The worst-affected village was Hafun, an island outcrop linked to the mainland by a single bridge. The beach there was littered with the debris of the tsunami: dozens of crumpled fishing boats, tangled piles of fishing nets, shoes and clothing. Locals said that many of those who died became caught up in the nets and were thus unable to swim free. Witnesses said the water rushed in and out four times. When the sea was sucked out on the first occasion, it exposed the lobster beds, and many villagers who ran to collect lobsters were stranded when the wave came back in with renewed vigour…

Chapter opener and above: 3,000 miles from the epicentre of the earthquake, the tsunami struck Hafun in north Somalia.

Mahado Musa's son died after she was forced to choose between him and his sister when the wave smashed into her home. Since six-year-old Mohammed was too heavy to carry, she picked up her two-year-old daughter, Khadra, and ran for her life. Her brother-in-law tried to save the boy but found him unconscious in the water, having probably been hurled against the walls by the waves. He died that night.

For the people of Somalia, the tsunami was the latest in a succession of disasters, each following on the heels of the one before. First, they had four consecutive years of drought; next there were heavy rains that killed many of the livestock which had survived the drought; and then the tsunami caused wholesale suffering.

The tsunami also struck further down the East African coast, killing ten people in Tanzania, one in Kenya (a man swimming in the sea), and three in the Seychelles where a bridge linking the main airport with the capital Victoria was destroyed. Over 1,000 people were made homeless on the island of Madagascar.

In contrast to Somalia, the remote Indian-ruled Andaman and Nicobar Islands were close to the earthquake's epicentre. The archipelago consists of 572 islands, 36 of them inhabited, and it was the more southerly Nicobar Islands that faced the full wrath of the tsunami. Many of the 20,000 residents of the low-lying Car Nicobar – a mixture of Indian migrants and tribal Nicobarese – were still asleep when the first

THE NET OF DESTRUCTION

waves came in. They emerged from their flimsy seaside houses to be confronted by a 15-metre (50 foot) high slab of water. Those who were quick escaped into the jungle; the slower ones were engulfed by the waves and dragged out into the Indian Ocean. "We ran up into the forest and hid," said Casper James, from the village of Malacca. "When we returned there was nothing left. I saw only hands sticking out of the sand."

Some inhabitants spent five days in the jungle with precious little to eat before eventually being rescued. They were so desperate for water that they dug holes in the ground before scooping up rainwater with their bare hands. One old man was seen climbing up a tamarind tree with his one-year-old grandson, but the tree was not tall enough and the waves swept them both away. Another woman began to run for the hills, only to realise that she had left her child behind. When she returned for the infant, both were carried out to sea where they drowned.

"We slept in the open," said one man. "We had no blanket, no carpet and no tent, and we were unable to open coconuts to drink because we had no knife. By the time help arrived we were very weak and hungry."

The tsunami made no distinction between rich and poor. The local magistrate was swept away, along with 68 Indian air force personnel and their families, plus tribal fishermen. One woman took her children to the fourth floor of one of the island's highest buildings but even there she was not safe as the giant waves ripped the infants from her arms. In a matter of minutes, half

A bridge linking the Seychelles international airport to the country's capital Victoria was snapped by the waves.

A member of the Indian airforce surveys the damage to an army residential complex at Car Nicobar Island, 2 January 2005.

of Car Nicobar's population – 10,000 – had vanished. Eighty per cent of the island was destroyed.

On Great Nicobar, one survivor said that after the second wave had smashed the family's concrete house, they cooked some dead fish brought in by the tsunami and then set off on a 20-kilometre (12 mile) trek through the crocodile-infested jungle.

At Indira Point on the Nicobar Islands, India's most southerly tip and just 160 kilometres (100 miles) from the earthquake's epicentre, four scientists who had been studying giant leatherback turtles were swept out to sea. The 30-metre-high lighthouse at the point was under water and the staff and their families had disap-

peared off the face of the earth.

Meanwhile on the tiny island of Chaura, PC Michael Paul was locking up his police station when a huge wave sent him crashing to the ground. "I had no chance to escape," he said, "because then a wall fell on top of me. The wave took me into the jungle, and then brought me back into the sea." Badly injured, he was eventually dragged by colleagues to higher ground where they spent two days waiting to be rescued, surviving on coconut juice and wild jungle potatoes. 65 per cent of the island was reported to be underwater in the wake of the tsunami while Pilmillow Island in South Nicobar was more or

THE NET OF DESTRUCTION

less totally submerged. On other islands such as Little Andaman, witnesses reported that crocodiles had started eating bodies that were floating in the water.

However, fears that the islands' almost extinct aboriginal tribes had been wiped out proved groundless. The Andaman and Nicobar Islands are home to five primitive tribes of Mongoloid and African origins, including the Sentinelese, a prehistoric tribe of hunter-gatherers who fiercely resist all contact with the outside world. These tribes originally migrated to the islands over 30,000 years ago, but their numbers have dwindled over the past two centuries. Because they inhabit the jungles on the highest points of the islands, they were spared the worst of the tsunami.

Successive Indian governments have offered incentives for the impoverished from the mainland to go and live on the islands. But having lost everything in the floods, many were eager to return home.

An indication of the force with which these tiny islands were battered by both the earthquake and the tsunami can be gauged from the fact that scientific data revealed the capital Port Blair to have shifted 1.15 metres away from mainland India as a result of events on 26 December.

Yet while the huge waves were destroying his parents' village of Hut Bay on Little Andaman, a baby boy was being born in upland woods. As the sea began to engulf their coastal settlement, Lakshmi Narain Roy grabbed his six-year-old son and pregnant wife Namita, put them in a rickshaw and pedalled for the hills. Mrs Roy briefly fell and lost consciousness and when she later complained of a pain in the abdomen, her husband attributed it to the fall since the baby was not due until 15 January. "But as the pain got worse by nightfall," said Mr Roy, "I became frantic and started

An aerial view of the Hafun coast where, locals say, many people died because they got tangled in fishing nets and were unable to swim free.

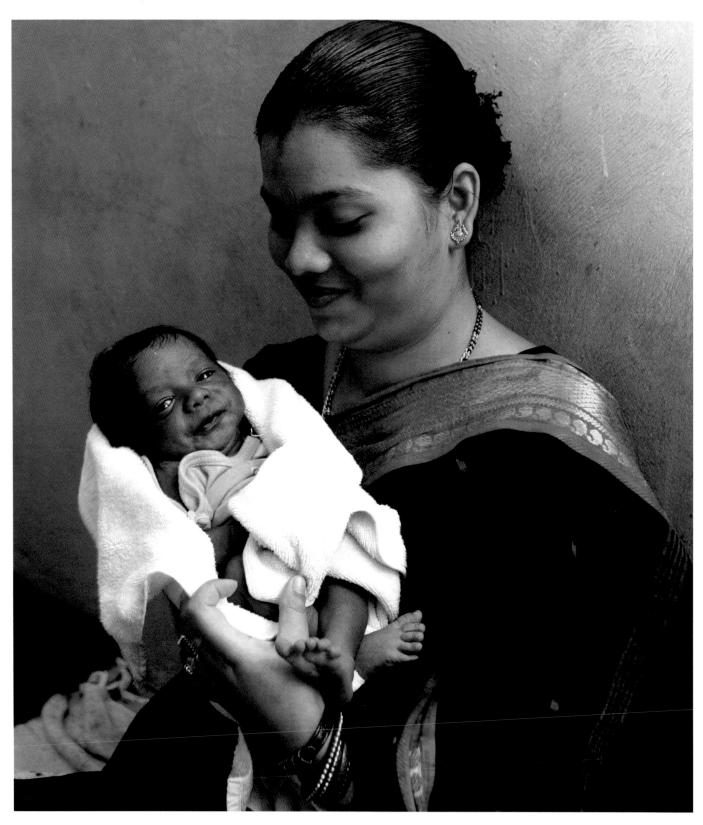

Miracle baby: Proud mother Namita Roy holds her son, Tsunami Shankar, born moments after the waves struck Andaman.

THE NET OF DESTRUCTION

Malaysian police assist with the evacuation of tsunami victims' bodies, 18 January 2005.

looking for help. Luckily I found a nurse." A makeshift curtain was erected, and the nurse and some other women helped Mrs Roy onto a bed of dried leaves and grass, and ordered the men to fetch clean cloth, thread and a bowl of hot water. After the baby was born Mrs Roy was still experiencing pain because the nurse, with no instruments, had been unable to remove the placenta from inside the mother's womb. Four days later, the family heard that a Royal Navy ship had arrived and so they travelled out to sea on a dinghy to get help. There then followed a seven-hour journey to a hospital in Port Blair where doctors were at last able to treat Mrs

Roy. At the suggestion of doctors, the miracle baby was named Tsunami. "After all," said Mr Roy, "it is a name everyone will instantly notice and remember."

The tsunami hit Malaysia shortly after 8am. On the northern island of Penang, children were happily playing on the beach, fascinated by the huge wave that was approaching. Out in the sea, people were swimming or riding on jet-skis. Canadian tourist Jasper Bintner, who was staying at Batu Ferringhi, north-west of Penang Island, said: "At first, you could just see a wall of waves in the distance with the white tops crashing down. Luckily, we had a lot of visual warning so we could get

Hindu devotees worship at the edge of the ocean on the holy day Makar Sankranti, Port Blair, 14 January 2005.

out of the water and the locals made sure we did. Around the corner, where the people were washed out to sea, they didn't have any warning. The tsunami just swept them off the beach."

Over 70 people were killed in total, the majority on Penang. There were also casualties on the mainland, where the earthquake damaged hundreds of buildings, but much of the coastline was spared because it was shielded from the tsunami by Sumatra.

Like Thailand, the Maldives – a cluster of 1,192 tiny coral islands in the Indian Ocean, off the south-west coast of India – is a favourite destination for western holidaymakers, with the palm trees, white sands and clear, turquoise blue seas representing many tourists' idea of paradise. Although the shallow waters and outlying coral reefs limited the tsunami's

destructive power, much of the islands' land mass barely rises above sea level, which left them exposed to widespread flooding. Two-thirds of the capital island, Male, was flooded, some of the city being saved by a sea wall, and outlying atolls were completely submerged. Twenty of the Maldives' 199 inhabited islands were described as "totally destroyed" while 53 suffered severe damage. Of the inhabited islands, 14 were completely evacuated, 79 were left without safe drinking water, 26 were left without electricity, and 24 had no telephones.

A British tourist died of a heart attack on seeing the rising water heading towards him at around 9.30 that morning and a total of at least 80 people were drowned. "The whole sea just lifted up," reported one witness in Male. "There was no sound. It just poured on to the

THE NET OF DESTRUCTION

Tsunami survivors from Car Nicobar and Campbell Bay queue for a midday meal at a relief camp in Port Blair, 13 January 2005.

island. Small boats were dropped on to the street."

One resourceful father tied his young children tightly to trees with beach towels to prevent the sea snatching them away.

While crops were lost as a result of salt-water contamination, the major damage in the Maldives was to the tourist industry. Many of the islands' luxury beachfront resorts faced the prospect of months of repairs before being able to reopen. Nevertheless, there was a sense of relief that these tiny islands had not suffered human losses on the same scale as, say, Sri Lanka. Not only did the coral reefs protect the islands but because there was no land mass for the swell of water to build up against, the waves that hit the Maldives did not exceed 1.5 metres (5 feet) in height.

Burma (with around 60 deaths) and Bangladesh (two) were also hit, but one place that somehow escaped the fury of the tsunami was the fishing village of Dhanushkodi on the eastern end of the island of Rameswaram. From there it is only 18 kilometres (11 miles) to Sri Lanka, yet whilst the latter saw much of its coastline devastated, Dhanushkodi remained untouched. The water simply rose a little and then went back. Even so, the local fishing industry was hit. For a few weeks after the tsunami, many fishermen from Dhanushkodi were afraid to venture out to sea and even those that did were unable to sell their catch amid public fears that it would be contaminated from all the dirt and dead bodies in the water. Sadly it was a problem that faced fishing communities right across the Indian Ocean, and was just another of the tsunami's terrible legacies.

THE REBUILDING 8
OPERATION

After the tsunami had destroyed cities, villages and industries in just a few minutes, came the long, slow process of attempting to repair the damage. Days of national mourning were declared in all the stricken countries, periods of silence were held across the globe, but the most pressing need was obviously the humanitarian crisis. Some of the world's poorest areas were among the worst affected, and it was immediately apparent that unless billions of dollars worth of aid was forthcoming, thousands more would die from disease and malnutrition in the weeks following the disaster.

Perhaps because so many nations far from the devastated area had lost people to the disaster – Sweden alone is believed to have lost several hundred tourists – governments dug deep to pledge financial aid. Corporations and companies followed suit. But it was the response from ordinary men, women and children that was truly staggering and which went some way to restoring at least a little faith in human nature...

As nightly news bulletins beamed over yet more distressing scenes from Banda Aceh, Phuket or Sri Lanka – to name but three – hundreds of thousands of people donated money to the tsunami appeal. They did it at supermarkets, banks, over the Internet, just about anywhere. Charity fund-raising events – big and small – were held across the globe, from humble car-boot sales to the concert at Cardiff's Millennium Stadium, featuring the likes of Eric Clapton, Keane and Snow Patrol in a 2005 version of Live Aid and raised over £1,300,000 for the tsunami appeal. Many celebrities have also pledged large sums of money, racing driver Michael Schumacher leading the way by donating £7 million. But all the money promised or donated in Britain, the United States, Germany or Australia would be worthless unless it was translated into solid aid on the ground in the places where it was needed most. The priority was to feed the survivors of the tsunami.

Chapter opener and above: A US Navy helicopter hovers near a village on the island of Sumatra and crewmen dispense much-needed food, water and relief supplies to Indonesian citizens, 7 January 2005.

However in some stricken areas this was easier said than done. Lines of communication – bridges and roads – had been swept away by the giant waves with the result that many towns and villages were only accessible by airplane or helicopter. In these places, regular food drops by relief organisations were the difference between life and death.

In India, aid agencies like World Vision, Oxfam and the Red Cross supplied rice, dhal, fish, flour, sugar and cereal as well as fresh water, blankets, plastic sheeting and other essential items such as soap, clothing, candles and matches. In Sri Lanka, where five per cent of the country's population was left homeless, thousands of Sri Lankans rushed to offer donations of food and clothing. Three-wheelers, flying the white flag of

Elephants assist the clear-up operation in Banda Aceh, 6 January 2005.

mourning, raced through Colombo with their hooters blaring, urging people to give and Buddhist monks drove pick-up trucks to collect supplies bought by well-wishers from supermarkets. Palm fronds were brought in to make roofs for temporary shelters. With thousands of people living in refugee camps across Asia, it was a question of tiding them over from one day to the next until some semblance of normality was able to return to their lives.

But very little was straightforward. Days after the tsunami, relief efforts in Sri Lanka were hampered by heavy monsoon rain which washed away roads and caused fresh flooding. And it emerged that the tsunami floodwater had uprooted land mines, thus threatening to kill or maim aid workers and survivors.

A sniffer dog searches for the bodies of tsunami victims in the Takuapa district, Thailand, 6 January 2005.

Thai rescue team members handle a dead body during search operations in the Phang Nga province, southern Thailand, 2 January 2005.

Meanwhile the task of ferrying aid to Indonesia was held up for several hours at one point when a cargo plane hit a herd of cows that had strayed on to the runway at Banda Aceh airport. With the runway temporarily out of action, hundreds of transport aircraft were unable to make essential deliveries. The delay in handing out food led to reports of looting and fights broke out on the streets over packets of noodles dropped from military vehicles. Hospitals were overflowing, while doctors complained that a shortage of vehicles for their medical teams was preventing them reaching refugee camps. Similarly, relief parcels piled up at airports in Medan and Banda Aceh because there not enough trucks to transport them. Although these airports were functioning, most of the smaller gravel airfields had been damaged by the earthquake and the tsunami, and because the majority of roads in the region were confined to the coast due to the rugged interior, the communication links were all but

The shocking human cost of the tsunami: a mass grave close to the Bang Mak Roang temple in Takua Pa, Thailand, 6 January 2005.

Indonesian policemen undergo the grim task of throwing bodies into a mass grave in Banda Aceh, 18 January 2005.

destroyed. Even 11 days after the disaster, only limited supplies had reached such devastated towns as Meulaboh. To make matters worse, two large aftershocks three days after the original earthquake sparked fresh panic among the Acehnese.

There were also problems with the Free Aceh Movement. Despite rebel leaders declaring a ceasefire, there were claims that insurgents had attacked Indonesian Army soldiers on humanitarian missions. It was a similar story in Sri Lanka where the north of the country was Tamil Tiger rebel territory, making it difficult for relief to get through.

Repairing the flood damage in strife-torn Somalia was made harder by the fact that the country lacked paved roads and adequate ports after 14 years without a central government. Consequently trucks carrying relief supplies faced a 464 kilometre (290 mile) journey along bumpy dirt tracks from the Red Sea port of Bossaso. And even those basic roads were obliterated near the coast, causing trucks loaded with food to be stuck in sands for hours – sometimes days – before they could reach their intended destination.

The Acehnese people divide meat portions of sacrificed animals for the poor during the second day of Eid-Al-Adha, the most important feast in the Muslim calendar, 22 January 2005.

There was widespread unrest on the Andaman and Nicobar Islands where local people felt they had been abandoned by the Indian government. Whilst the government blamed the delay in getting relief to the islands on their geographical remoteness and the fact that pontoons and jetties had been washed away, making it difficult for rescue boats to land, aid workers in Port Blair were highly critical. "Dead bodies have just been left here," said one. "We should be angry with this administration. They were slow to rescue people, they have given us no help. We have been given no medicines."

Perhaps the greatest threat, however, was from disease. With so much contaminated drinking water and so many rotting corpses lying around, the danger from cholera, diphtheria, typhoid, diarrhoea, hepatitis and other diseases was all too real. As a result piles of unclaimed bodies were buried prematurely in mass graves in an attempt to prevent the spread of disease. In Sri Lanka's devastated town of Galle, officials mounted a loudspeaker on a fire engine to ask residents to lay dead bodies on roads ready for collection and burial. Elsewhere in the country, residents buried corpses with forks or even

An aerial view of one of the temporary relief villages set up in Banda Aceh to offer shelter to Indonesians left homeless by the tsunami, 12 January 2005.

their bare hands. In Banda Aceh, 20,000 children were vaccinated against measles but plans to vaccinate a million others were hampered by a lack of qualified medical staff, many of whom were killed in the disaster.

The Red Cross revealed: "The biggest health challenge we are facing is the spread of water-borne diseases, particularly malaria and diarrhoea, as well as respiratory tract infections."

Many individuals were not content with simply donating money. Touched by the suffering, they wanted to offer practical help and caught

13-year-old Nadika, receives treatment from Alison Thompson, a nurse from New York, working as a volunteer in a field hospital in the village of Paraliya, southern Sri Lanka, 13 January 2005.

the first available flight out to places such as Banda Aceh, Galle and Phuket where they offered their services as required. Some went out with the intention of working as builders or carpenters, helping to reconstruct damaged homes, only to find themselves pressed into action as temporary morgue porters, carrying hundreds of bodies from one place to another.

According to United Nations figures, over one million people lost their jobs as a result of the disaster. Unemployment in the hardest-hit areas of Indonesia was put at over 30 per cent, compared to just 6.8 per cent before the tsunami. In Sri Lanka, more than 400,000 workers along the coastline lost their jobs, most of them in fishing and tourism.

The priority for the Thai authorities was to restore confidence in the

South Korean Red Cross volunteers in Suwon prepare emergency relief kits to be sent to Indonesia, 28 December 2004.

'After the tsunami
had destroyed
cities, villages and
industries in just a
few minutes, came
the long slow process
of attempting to
repair the damage'

country's tourist industry. With 12 million people visiting Thailand every year, the nation was heavily dependent on tourism, but in the wake of the tsunami it was estimated that 200,000 employees in the tourism sector would lose their jobs. If holidaymakers failed to return, the entire Thai economy – from beachfront bars and watersports rental companies to top hotels and airlines – would be badly hit. Anxious to present a hint of 'business as usual', by 12 January some of the affected resorts had reopened and the government had launched an advertising campaign to bring visitors back to the area as quickly as possible. Ironically, in the long term the disaster may benefit Thailand as money pledged by overseas governments will be spent partly on building superior hotels to replace the flimsy structures destroyed by the tsunami.

Another area heavily reliant on tourism – the Maldives – suffered an estimated $4.8 billion of damage. Government spokesman Ahmed Shaheed said: "The tsunami has within a few minutes set the country back by at least two decades as far as socio-economic development is concerned." Nineteen of the country's 87 luxury resorts were badly damaged and were

Jenny Behlin-Swedel from Stockholm, Sweden, a Tsunami survivor and relief volunteer, updates the missing persons list at a hospital in Phuket, 4 January 2005.

Survivors from Nicobar island sit together at a tsunami relief camp in Port Blair on New Year's Day 2005.

Sri Lankan women walk with
their children at a relief camp in
Kattankudi, 6 January 2005.

forced to close for several months. Inevitably bookings suffered, dropping
by half in what was the peak season.

Throughout the region there was a public aversion to eating locally-
caught fish for fear that the fish had fed on human corpses swept out to sea
by the tsunami. This had a crippling effect on local fishing industries with
distributors refusing to buy produce from affected areas. For example,
rather than buy from the Andaman Sea ports in Thailand, they preferred to
do business in Malaysia or Vietnam so that they could assure customers
there was no risk of contamination.

The earthquake and tsunami also proved to be an environmental disaster
that will affect the Indian Ocean region for years to come. Mangroves,

coral reefs, forests, coastal wetlands, vegetation, sand dunes and rock formations were all seriously damaged while the destruction of sewage collectors and treatment plants pose an additional threat. Thousands of rice, mango and banana plantations in Sri Lanka were destroyed.

As governments began the cleaning-up operation, they were left to reflect that so many human lives could have been saved had a tsunami early warning system been in place in the Indian Ocean, as it is in the Pacific. Professor Bill McGuire of the Benfield Hazard Research Centre declared: "At least two-thirds of the people who died should not have died. With a warning system they could have had an hour or so to get a kilometre or two inland or to reach high ground."

With Thailand, in particular, backing such a system to restore tourists' confidence, the clamour became deafening and three weeks after the

A shaft of sunlight illuminates rows of coffins inside a warehouse at Phuket port, 28 December 2004.

American Navy personnel aboard the USS Abraham Lincoln fill containers with purified water to be sent to Sumatra, 4 January 2005.

THE REBUILDING OPERATION

disaster, delegates at a United Nations-sponsored conference in Kobe, Japan, pledged to establish a tsunami warning system in the Indian Ocean by the middle of 2006. Kofi Annan, secretary general of the UN, said: "We must draw on every lesson we can to prevent tragedies like this occurring in the future. Prevention and early warning systems must become a priority."

Whilst experts welcomed the news, they stressed that educating the public about tsunamis was equally important.

Things would never be the same again in the countries bordering the Indian Ocean, but at least it was a start.

Fundraising events were staged all over the world to aid tsunami vicitims, from car-boot sales to a pop concert at the Millenium Stadium in Wales that raised over £1.3 million.

TSUMANIS 9
IN WAITING

Following the horror of the 2004 Indian Ocean tsunami, the question on the lips of geologists across the world was: where next? Whilst undersea earthquakes are difficult to predict, unstable volcanoes invariably offer more tangible warning signs, bubbling and rumbling away for years before finally erupting. And when a volcanic eruption coincides with the threat of a massive landslip into the ocean, the potential for a devastating tsunami is all too obvious. Thus western scientists searching for the "next big thing" did not have far to look: they simply focused on the island of La Palma in the Canaries.

Situated off the west coast of Africa, the Canary Islands are an archipelago of seven big islands and several small islets, all formed by volcanoes on the seabed. The chain of islands formed as the Earth's crust moved across an upwelling of molten rock from deep below the planet's surface and, from the evidence of abundant landslide deposits strewn about their bases, the Canary Island volcanoes have experienced at least a dozen major collapses…

CHAPTER OPENER: A view into La Palma's Caldera de Taburiente, one of the world's largest erosion basins at 5 kilometres wide and two kilometres deep. ABOVE: A map showing the potential reach of a mega-tsunami caused by a landslide at La Palma's Cumbre Veija volcano.

over the past several million years. Although La Palma's volcano, the Cumbre Vieja, has not erupted since 1971, it remains active, but it is the island's history and contours that lead scientists to fear the worst.

With a peak of 2,426 metres (7,900 feet), the steepness of La Palma means that there is always a danger of unstable land tumbling into the sea below. When the Cumbre Vieja erupted in 1949, a large chunk of its western flank slipped four metres (13 feet) towards the ocean. Thankfully, it stopped short of the water, but with the entire flank believed to be unstable, geologists think it is only a matter of time before the same thing happens again on a grander scale…and with more catastrophic consequences.

Dr Simon Day of the Benfield Hazard Research Centre at University College, London, and Dr Steven N. Ward of the University of California fear that any future eruption of the Cumbre Vieja could see an area of rock the size of the Isle of Man crash into the sea, displacing enough water in the Atlantic Ocean to send killer tsunamis racing to four continents. The main volcanic fissure on the Cumbre Vieja runs north-south along the ridge. During any future eruption, hot rock would be forced upwards, widening the existing cracks that separate the western flank from the ridge. The build-up of volcanic gases and the pressure from hot groundwater may force the cracks open further still, pushing the flank completely away from the ridge, and causing the collapse of a mass of rock weighing half a trillion tonnes. This land mass could slide into the ocean in just 90 seconds. If, in the worst-case scenario, it stayed in one piece as it hit the water and slid along the deep seabed, it could generate an initial wave a colossal 900 metres (3,000 feet) high. Reducing in height to 300 metres (1,000 feet) as they surged across the ocean at speeds of 900 km/hour (560mph), the walls of water would crash across neighbouring islands within 15 minutes. By 30 minutes, 100-metre-high (300 feet) waves would strike the holiday island of Tenerife. An hour after the collapse, waves of a similar height could crash onto the coasts of West Africa, Lanzarote and Fuerteventura. After three hours, five-metre-high (16 feet) waves would pound the coast of Spain, but it is on the other side of the Atlantic that people may have most to fear. For within six hours of a La Palma collapse, the eastern seaboard of the United States would be lashed by ten-metre-high (33 feet) high waves, sweeping away everything and everyone in their path up to 20 kilometres (12 miles) inland. Geologists calculate that the city of Boston would be hit first, followed by New York, then all the way down the coast to Miami and the Caribbean. Although Florida would be struck nine hours after the landfall, thus allowing plenty of time for warnings to be passed on to coastal areas, the height of the waves anticipated could be up to 25 metres (80 feet). Quite simply, people thousands of miles away who have never even heard of La Palma could be in grave danger.

South America would also be on the receiving end of waves up to 15 metres (50 feet) high, while Britain and Ireland could expect five-metre-high (16 feet) waves four or five hours after the collapse, causing damage to seaside ports and resorts. The reason for Europe experiencing smaller waves than the other continents is that La Palma Island itself would block most of the tsunami radiation in a northerly direction.

Of course there could be many more summit eruptions of the Cumbre Vieja before the western flank collapses.

A space shuttle image of the volcanic island at La Palma which is causing great concern among geologists.

Professor Bill McGuire, of the Benfield Hazard Research Centre, says: "There could be ten or there could be 20 – we simply don't know. But if I was living in Miami or New York and I heard that the Cumbre Vieja was erupting, I would keep a very close eye on the news."

So although the 2004 tsunami was caused by an earthquake, experts believe that landslides can create even bigger tsunamis. They cite the largest recorded wave in history, in Lituya Bay, Alaska, in 1958, which was caused by the collapse of a towering cliff. The resulting wave was higher than any skyscraper on Earth. The true destructive potential of landslide-generated tsunami, which scientists have named "mega-tsunami", suddenly began to be appreciated. If a modest-sized landslide in Alaska could create such a monster wave, what havoc could a really huge landslide cause?

Scientists scouring the world for potential danger zones have discovered that in the past giant landslides have taken place on the Cape Verde Islands, off the west coast of Africa, and on the island of Réunion in the Indian Ocean. Both locations are therefore considered a risk. Further east, although the 1883 eruption virtually destroyed Krakatoa, a new volcano called Anak Krakatoa ("Child of Krakatoa") replaced it in 1927. Growing at the rate of 4.5 metres (15 feet) per year, Anak Krakatoa is now almost as big as the original Krakatoa was in 1883. It is active and experts fear it could be preparing for another eruption.

Meanwhile on Hawaii the Kilauea volcano, one of the most active in the world, has been giving cause for concern. In what is known as a 'silent earthquake', a 72-square-mile chunk of the southern slope of the volcano slipped nine centimetres (3.5 inches) towards the sea some time in the late 1990s, leading scientists to warn of a possible disaster for Pacific Rim nations. If a land mass that size slid into the ocean, it could trigger an enormous tsunami that would devastate the most densely populated parts of Hawaii in just 20 minutes before smashing into Ecuador and imperilling coastlines as far away as California, Chile, Japan and Australia.

Seismic activity is monitored 24 hours a day at Japan's Meteorological Agency in Tokyo.

The last mega-tsunami occurred 4,000 years ago at the island of Réunion (pictured) in the Indian Ocean.

Happily, huge landslides and the ensuing mega-tsunami are extremely rare. The last one happened 4,000 years ago on the island of Réunion, but that does not mean the danger from La Palma should be dismissed. As Professor McGuire says: "The thing about La Palma is we know it's on the move now. It is not a matter of if the rock plunges into the ocean, it is a case of when."

Yet despite the potential scale of the threat, little is being done to monitor the geological activity of La Palma. Unlike Kilauea, which is bristling with hi-tech equipment installed by geologists to detect changes in the shape of the mountain, only a few seismometers have been set up on the precarious western flank of the Cumbre Vieja, and these will not provide sufficient information to predict when another eruption might take place. "It's a worrying situation," adds Professor McGuire. "It will almost certainly go during an eruption. The problem is that with just a few seismometers on

the island, we may not get the notice we need. We must have better monitoring so we know when an eruption is about to happen."

The cost of installing more sophisticated sensors, which could give two weeks' warning of an eruption, plus global positioning satellite units to detect how quickly the land mass was falling into the ocean, would be no more than a few hundred thousand dollars.

Other proposed solutions would have little effect. Coastal barriers would not be able to sustain the battering from a mega-tsunami, and breaking the island of La Palma apart before it collapses has been decreed too dangerous.

A La Palma tsunami could affect 100 million people across the world. The only certainty, it seems, is that at some point in the next few thousand years a huge section of the island will fall into the Atlantic Ocean. It is the responsibility of the world's leaders to be prepared.

Photographic Credits

The publishers would like to thanks the following sources for their kind permission to reproduce the pictures in this book. The page numbers for each of the photographs are listed below, giving the page on which they appear in the book. Any location indicator (c-centre, t-top, b-bottom, r-right, l-left).

Sources of Information

www.guardian.co.uk www.news.bbc.co.uk www.en.wikipedia.org www.us.rediff.com www.news.nationalgeographic.com www.livescience.com www.abc.net
Honolulu Advertiser Honolulu Star-Bulletin New York Post